Steven Weinberg: A Life in Physics

Steven Weinberg shares his candid thoughts, in his own words, on theoretical physics and cosmology, along with personal anecdotes and recollections of the people who helped shape his career. These memoirs of his life as a scientist and public figure cover his student days and early career, through the golden age of particle physics in the 1970s, his being awarded the Nobel Prize, through to the end of the twentieth century. In addition to his research insights, Weinberg provides glimpses into his life in academia more broadly: dealing with the "two-body problem," tenure, international conference travel, his book-writing, advisory work with JASON, and his advocacy for the Superconducting Super Collider. Physicists, historians of science, and interested readers will find the presentation engaging and often witty, as Weinberg reflects on his life in physics.

STEVEN WEINBERG (1933–2021) held the Josey Regental Chair in Science at the University of Texas at Austin. A Nobel laureate in physics and National Medal of Science winner, he is recognized as one of the key architects of the Standard Model of particle physics. In addition to his ground-breaking research in fundamental physics and his science advocacy, he was a prolific author of both academic and popular-level books, widely celebrated for his exceptional physical insight and his gift for clear exposition.

Steven Weinberg: A Life in Physics

STEVEN WEINBERG
University of Texas at Austin

Shaftesbury Road, Cambridge CB2 8EA, United Kingdom

One Liberty Plaza, 20th Floor, New York, NY 10006, USA

477 Williamstown Road, Port Melbourne, VIC 3207, Australia

314–321, 3rd Floor, Plot 3, Splendor Forum, Jasola District Centre, New Delhi – 110025, India

103 Penang Road, #05–06/07, Visioncrest Commercial, Singapore 238467

Cambridge University Press is part of Cambridge University Press & Assessment, a department of the University of Cambridge.

We share the University's mission to contribute to society through the pursuit of education, learning and research at the highest international levels of excellence.

www.cambridge.org
Information on this title: www.cambridge.org/9781009513470

DOI: 10.1017/9781009513487

© The Estate of Steven Weinberg 2025

This publication is in copyright. Subject to statutory exception and to the provisions of relevant collective licensing agreements, no reproduction of any part may take place without the written permission of Cambridge University Press & Assessment.

When citing this work, please include a reference to the DOI 10.1017/9781009513487

First published 2025

Printed in the United Kingdom by CPI Group Ltd, Croydon CR0 4YY

A catalogue record for this publication is available from the British Library.

Library of Congress Cataloging-in-Publication Data
Names: Weinberg, Steven, 1933–2021, author.
Title: Steven Weinberg : life in physics / Steven Weinberg, University of Texas, Austin.
Other titles: Steven Weinberg, life in physics
Description: Cambridge, United Kingdom ; New York, NY, USA : Cambridge University Press, 2025. | Includes bibliographical references and index.
Identifiers: LCCN 2024011926 | ISBN 9781009513470 (hardback) | ISBN 9781009513487 (ebook)
Subjects: LCSH: Weinberg, Steven, 1933-2021. | Physicists – United States – Biography. | Nobel Prize winners – United States – Biography. | Austin (Texas) – Biography.
Classification: LCC QC16.W45 A3 2025 | DDC 530/.092 [B]–dc23/eng/20240508
LC record available at https://lccn.loc.gov/2024011926

ISBN 978-1-009-51347-0 Hardback

Cambridge University Press & Assessment has no responsibility for the persistence or accuracy of URLs for external or third-party internet websites referred to in this publication and does not guarantee that any content on such websites is, or will remain, accurate or appropriate.

Contents

Preface: The Twentieth Century	*page*	vii
Publisher's Note		viii
1 *First Things*		1
2 *Turning to Science*		3
3 *Cornell*		8
4 *Copenhagen*		18
5 *Princeton*		24
6 *Manhattan*		35
7 *San Francisco and Berkeley*		45
8 *East to London*		60
9 *Berkeley*		72
10 *Cambridge: 1966–69*		93
11 *Cambridge: 1969–72*		112
12 *Cambridge: 1972–79*		125
13 *Gone to Texas*		173
14 *Super Collider Days*		210
15 *Austin: The 1980s*		227
16 *The Dark Energy*		232

17 *Austin: The 1990s* 237

    Image Credits 245
    Bibliography 247
    Index 249

Preface: The Twentieth Century

In this new millennium, we are witnessing a golden age of cosmology, and I have greatly enjoyed working at its dawn in the twentieth century. I will be turning very soon to a second volume of these memoirs, which will concern my work in twenty-first-century physics and cosmology. Here, I have tried to convey the excitement of the advances we made only yesterday, in the twentieth century.

In my life, I had at least a casual acquaintance with some of the leading physicists of the twentieth century: Bohr, Dirac, Dyson, Feynman, Hoyle, Oppenheimer, Pauli, and Schwinger. I worked with a brilliant later generation of physicists to build the Standard Model of elementary particle physics. At a very low level, I participated in public controversy over ballistic missile defense, the manned space flight program, and the ill-fated Superconducting SuperCollider project.

Now, in the new millennium, so much is opening up to us! We are explaining the dark energy that permeates the universe. We are trying to understand the dark matter within it. Now the Hubble Space Telescope is seeking out the beginnings of time and space, and sending messages back to us. And the James Webb Space Telescope will soon be launched, orbiting the Sun as a planet does. The birth of the universe, when matter begins to take shape, will be the ultimate physics laboratory. How marvelous it is to be alive and in the thick of things now! But it was marvelous then too, in the twentieth century.

Steven Weinberg
Austin, June 2, 2021

Publisher's Note

At the time of his death on July 23, 2021, Steven Weinberg's memoirs were incomplete, covering his life and work up to the end of the twentieth century. The raw notes were collated by his widow, Louise Weinberg, Professor of Law Emeritus at the University of Texas at Austin. Following his instructions, Louise first edited them to remove any material that he would have wanted to edit out himself. This short volume therefore focuses primarily on his life as a scientist and public intellectual. Rather than attempting to complete or polish his writing, we have purposely left the chapters essentially as they were passed to us, other than to correct simple typos and the like. No doubt these memoirs would have been developed more fully, had he been granted the time.

Readers whose appetites have been whetted by these writings may wish to look out for his authorized biography by science writer Graham Farmelo, currently in preparation. For historians and researchers, Weinberg's papers are held at the Harry Ransom Center, The University of Texas at Austin; a detailed inventory may be viewed online.

We are grateful to Louise Weinberg for her permission to publish these memoirs, and for putting her trust in Cambridge University Press to take care of this legacy project, as a lasting record for future generations. We thank the handful of Weinberg's colleagues and collaborators who helpfully read and commented on the initial draft typescript.

The raw typescript was not illustrated, so we took the liberty of adding some illustrative photos of some of the people and places mentioned, aiming to portray them at around the same time in history as the text. We acknowledge the organizations and individuals who granted permission to publish photographs, especially the American Institute of Physics' Emilio Segrè Visual Archives, and Louise Weinberg for sharing some of her own images.

I   First Things

Whatever native intelligence and intellectual curiosity I may have, I owe to my parents, and, in particular, my father. In my memories, his own native intelligence is becoming more and more apparent to me.

My grandfather, a Romanian immigrant, had risen to the middle class as a linguist in the civil service of the municipal courts in New York. He had been revered among the city's judiciary for his ability to translate the testimony and papers of other immigrants in myriad European languages.

One of his children, my father, born in America, fell in love with a beautiful, chic, quick-witted city girl from Berlin, Eva Israel, my mother. She had come to New York with her sister in 1928. Unable to fund both a family and a higher education, and fearfully insecure in the Great Depression, my father clung to civil service within New York's judicial system, as had his linguist father before him. My father, like his, also became famous among local trial judges, in his case, as the fastest court stenographer in the city, even after machine stenography came in. Judges would ask for him, as they had asked for his father.

He was determined that I should have the chances that had been out of reach for him. I remember the heterogeneous books he brought home and studiously did not bring to my attention. He bought used books literally by the bagful, and treated them as if bought for himself. This bulging library of miscellany was an exciting feature of my boyhood. I think my father understood that books in themselves were an education, and needed to be available to his son – and that the more I read, of anything, the better off I would be.

One of these books had a great impact on me. It was the dawn of my interest in cosmology. It came to me in a curiously primitive but beautiful vision of the creation of the universe, the Kearys' *The Heroes of Asgard*. This was a retelling of Norse myths from the Poetic Edda, written in medieval Iceland, and also from Saemund's Edda. The book was not just a collection of myths; rather, it anthologized fine translations from the Old Norse. I found the poetry arresting and the mystery transforming. Here are lines from the Edda's story of creation:

> Once was the age
> When all was not –
> No sand, nor sea,
> No salt waves,
> No earth was found,
> Nor over-skies,
> But yawning precipice
> And nowhere grass.

When I wrote *The First Three Minutes* (1977), my book on modern cosmology for the general reader, I began it with the Edda's cosmogony. Near the end, describing the idea of an oscillating universe, I referred to the story of Ragnarok, the story of the end of the world. There it is predicted that, afterward, the sons of Thor would come up from Hell carrying their father's hammer, and the whole story would begin again. Today, in a similar spirit, with the suggestion of late-twentieth-century string theory in physics, we can conceive of the strange possibility of a multiverse.

2 Turning to Science

From my days with a child's Chemcraft chemistry set, I had picked up bits of knowledge of chemistry. It was exciting to know the elements, the fundamental ingredients of which all matter is made.

I had learned that an element, such as gold, was made of atoms of a particular kind. Compounds of elements were made of molecules consisting of definite numbers of various atoms. For example, a molecule of water is compounded of two atoms of hydrogen and one of oxygen. That is why water is denoted as H_2O.

I was surprised to learn that atoms of different elements were described by a positive or negative whole number called their valence, and that stable molecules tended to be those whose atomic valences added up to zero. I did not know why, but I gathered that the explanation was offered by a science underlying chemistry, the science of physics.

Having perceived that physics was at the bottom of it, I wanted to find popular books on physics to clue me in. I tried to choose books by serious physicists. This did not make everything clear for me – just the opposite. What did become clear was that atomic physics was too mysterious for me to understand. That in itself was interesting: a mystery I could try to solve.

I read a series of works by the nuclear physicist George Gamow. (I later learned that Gamow had worked out a theory explaining a kind of radioactivity. Gamow was one of the first to apply physics to his thinking about the early universe.) The popular Gamow books I was reading followed the adventures of Mr. Tompkins, who found himself in worlds in which he encountered large weird effects, described by special relativity and quantum mechanics, effects that in the real world are normally too tiny to be observed. They show up more clearly at very high speeds approaching the speed of light, and at very small distances approaching the size of atoms.

I also read Gamow's *One, Two, Three, … Infinity*, which revealed that there were different kinds of infinity, and his *The Birth and Death of the Sun*, which described what stars are like inside.

But the astrophysicist Sir James Jeans had the biggest impact on me with his book *The Mysterious Universe*. There I saw an equation having something to do with the impossibility of simultaneous measurement of the momentum and position of any particle. Jeans argued that position times momentum would not equal momentum times position. I saw that these phenomena depended on time and each other. How then to deal with them? You would need special mathematics.

There was one book called *Heat*, the author of which I never found and which I never read but only saw once, lying open on a table in the public library, showing a symbol like a tall skinny S with a circle through its middle. I did not know what it meant, but I knew that symbols like that filled books on higher mathematics. So something as familiar as heat could be described by higher mathematics! I felt like Goethe's "Faust," opening a book and seeing the sign of the macrocosm: "Each character on which my eye reposes, Nature in act before my soul discloses." I felt I had to learn these kinds of mathematics.

Science fiction also stirred my unschooled imagination. I read classics by Jules Verne, H. G. Wells, and Olaf Stapledon, and also more recent books and stories by Isaac Asimov, Arthur C. Clarke, Robert Heinlein, Theodore Sturgeon, and A. E. Van Vogt. For good or evil, scientists figured in much of this fiction as figures of power.

Beyond science fiction, while at a summer camp in 1945 I had no idea what an atomic bomb was, by the time I entered high school three years later, everyone knew what an atomic bomb was, and it had become widely known that scientists had played a major part in the fight in World War II. The charismatic central figure was Robert Oppenheimer. In a 2005 review of a book about Oppenheimer,

I reminisced about how I was affected by his example. I had seen an issue of some magazine, probably *Life*, with Oppenheimer on the cover. I had read that Oppenheimer engaged in in-depth research on elementary particles and collapsing stars, that he was a student of Sanskrit poetry and an accomplished horseman, and, at the same time, that he had done much to help win World War II through his leadership at Los Alamos. Wow! I saw that you didn't have to renounce the world when you took on the vocation of theoretical physics. If to be a physicist was to be a priest celebrating some arcane mystery, still, it was possible to be a worldly one.

By January 1948, I had decided to become a physicist and was excited to start this process, having been admitted to the famous Bronx High School of Science.

The special feature of the Bronx Science curriculum was that it offered a variety of advanced science and mathematics courses. Alas, these courses had outlived their usefulness, in the sense that they were not what one meant today when one referred to that discipline. For example, if you wanted to take a course in "physics," you had a choice between automotive engineering and radio engineering. In chemistry, there were courses in laboratory techniques. Advanced mathematics was solid geometry.

FIGURE 2.1 The original Bronx High School of Science logo

The most serious insufficiency was that the school offered no course in calculus. That was true in most high schools, but a high school purporting to specialize in science might have been expected to do better. It may simply have reflected a changed job market for people who knew calculus. In later life, I learned that great universities, while offering calculus, did not require calculus for graduation, much less for entry. Yet calculus is a golden key that opens many doors in the mind and many opportunities in life, from architecture to engineering to the physical sciences. It could well be substituted in high schools for courses in trigonometry, a mathematics of limited use.

The true glory of Bronx Science is its student body, and the atmosphere of self-teaching the students create. The most important educational experience I had in high school was teaching myself calculus. Serious students at Bronx Science had no choice, as far as calculus was concerned. It is the language of physics.

Calculus was one of the great epiphanies of my life. I saw that with calculus I was able to calculate the orbits of planets, or the shape of the cables in a suspension bridge that would support a uniform horizontal load. It was like finding a bottle on a beach that, when rubbed, would summon a genie.

My best friends at Bronx Science were two other future physicists, Gary (Gerald) Feinberg and Shelly (Sheldon) Glashow. I will be referring to them in parts of this book to come; our trajectories were to intersect many times in the years after high school. Of the three of us, I would say that Gary had the greatest raw intelligence. Shelly was the most urbane. We *Drei Kameraden* shared ideas about things we were reading and, at the same time, competed with each other. Of course all three of us were teaching ourselves calculus, but this was taken for granted. I quietly began to learn about differential equations, and I suspect that Gary and Shelly were doing so too.

I was taking summer jobs in my high school years, to save for college tuition. One evening at a summer camp where I was working, I hitched a ride and saw one of the greatest movies ever made, *The Red Shoes*. The movie had an enormous impact on me. For me, the essential feature of the movie was that the characters were a team, working toward a common goal, the creation of something new, to which each brought a special expertise. Despite conflicts, they recognized and respected each other's talents. In my life, my interactions with other scientists could give me the same feeling. Most of my work has been done alone, but sometimes when talking about physics with colleagues I find it gratifying that we share a common language and common goals, and I think of the teamwork in *The Red Shoes*. It was in a similar spirit that the contributions of physicists the world over worked in the twentieth century to develop what is today called "The Standard Model."

I won a New York State scholarship that made Cornell the most attractive choice of university for me, if I wanted to see something of the world outside New York City. This was also the choice for my friend Shelly Glashow. Cornell's physics department had Hans Bethe, whom I knew to be the first to work out the nuclear reactions that give energy to the Sun, so it had to be a pretty good place to study physics. Shelly made the same calculation. My other close friend, Gary Feinberg, could not imagine life outside New York, and chose Columbia.

Years later, in September 2009, as students throughout America were entering college, the *New York Times* invited me, along with a few public intellectuals, to contribute to the *Times*' op-ed page our thoughts about the undergraduate experience. I reminisced about my anticipations of college in the summer of 1950: "The summer of 1950, before I went to Cornell, I saw a Cornell course catalog. Reading it ... was for me like the experience of a starving man reading the menu of a good restaurant. The physics department had courses given by famous physicists. The mathematics department was offering a course on Hilbert Space." I would come to love Cornell, and there I did learn about Hilbert space.

3   Cornell

In my first year at Cornell, I was initiated into a tradition of physics about which controversy still lingers. There is a view of the history of science associated particularly with the name of Thomas Kuhn, which sees this history as a sequence of revolutions "paradigm shifts." In Kuhn's view, at each shift of paradigm, our understanding of nature changes so radically that we can no longer understand the science of the past.

This view is plainly wrong. It is uncomprehending. Fortunately, it is not mirrored in the education of physicists, at Cornell or anywhere else. It is true that physics changed radically in the twentieth century with the advent of relativity and quantum mechanics. We now know that the classical nonrelativistic physics of the nineteenth century is not valid at very small scales, or very high velocities, or in the neighborhood of very large masses. But relativity and quantum mechanics were developed on the foundation of classical nonrelativistic physics. That foundation must be learned by all students who aim to become physicists. We do not dismiss the great advances of the past; we build on them. They were all quite right, the great physicists of the past. Copernicus and Galileo were right. Newton and Maxwell and Einstein were right. Each made a great advance, and they were all right. Succeeding generations try to explain why, and our explanations change and improve.

At Cornell, when I took the standard first-year courses for physics students, I was studying mechanics, heat, light, and electromagnetism. The physics I learned in my first year at Cornell was not the fancy stuff, not quantum mechanics, not relativity, not the advances that I had been reading about in popular books, and that I longed so to understand. Most of what I was learning in these first-year courses at Cornell would have been familiar to any physicist at the end of the nineteenth century. But it was the essential foundation for everything else I would learn in physics, first steps in the process of turning me into a physicist. Kuhn was wrong because he did not understand that we progress. We build on what we know. The triumphs of the past are part of our equipment today. We actually do, as the saying goes, stand on the shoulders of giants.

Cornell had other satisfying offerings for me. From the beginning in mathematics I found I could go straight into advanced calculus. Now, for the first time in years, I was accessing mathematics I did not already know without having to teach it to myself.

I was rooming with a physics student, Danny Kleitman, who would later become a mathematician. I was invited to join Telluride House, a fraternity. It was in a richly funded building with large public rooms, a fireplace, and served meals. Telluride was not concerned with religions or nationalities, but rather with intellect and achievement. It included a few graduate students along with undergraduates, and even one or two faculty members. The year before I arrived at Cornell, Richard Feynman, the great physicist, had been a resident faculty member at Telluride, trailing behind him a cloud of Feynman anecdotes.

In my sophomore year, something happened to me that would profoundly affect my life in science. Whether because of suffering over my girl's previous involvement with another boy (the girl was Louise Goldwasser, whom I married) or dislike of the earnestness of the Telluride mystique, or disillusion with philosophy, for which I had lost respect, or anger at having to serve in the Reserve Officers' Training Corps, or all of them together, I went into a classic sophomore slump. I let all my classwork slide (except, of course, in physics and mathematics).

It was a blue funk. My grades sank. It was no surprise when, near the end of the spring 1952 term, I was told that I would not be offered "preferment" at Telluride for the coming academic year.

FIGURE 3.1 A modern view of Telluride House

Starting my junior year at Cornell, I was sharing the rent of an apartment with Shelly Glashow. We got on each other's nerves, however, and I decamped to a cheap room. It had not helped that Shelly was capable of a certain devil-may-care irresponsibility. One weekend, having offered to drive Louise and me home to New York, he never showed up, but left us standing in the street with our suitcases.

As the rest of my junior year opened before me, I felt a great lifting of my spirits. This was not only because I had managed to disentangle myself from Glashow. Like a reformed drunk, I looked back in dismay at the harm I had done myself in my sophomore slump.

Something changed inside me at a deep level. I became happily incapable of wasting time. This happiness, when every moment is of use, is still with me. I feel uneasy at a waste of time. My sophomore slump had bestowed on me a lifetime given over to steady work.

There were two other oddities that made a difference in my work habits in science. First, once married to my girl I wanted to be

FIGURE 3.2 Portrait of the author, aged nineteen, by Louise Weinberg

with her all the time. When she was raising our daughter, I had to be working at home to be with her as much as I could, and the habit persisted even as her career bloomed and when she became a well-known professor of law. I would accompany her on her travels when she was invited to deliver distinguished named lectures or chair panels for the Association of American Law Schools meetings in New York or San Francisco, or Washington. I accompanied her to Philadelphia where she attended advisory council meetings at the lawyers' distinguished academy, the American Law Institute, of which she was a member. Philadelphia was a particular treat for me, because my old friend and former neighbor, Gino Segrè, was there at Penn. Similarly, she accompanied me on most of my travels, when she could get away.

Louise and I have always had our main offices at home together, two offices in every house we have lived in. It was one of the keys, I found, to productivity, to have my girl by my side.

There is another contribution to my productivity. While sitting at my desk at home doing physics or preparing classes, or doing some science writing, I picked up the habit of watching classic movies or the History Channel on television. My TV is always turned on in its corner of my desk. Doing the two things at once doubles the value of time. And the movie keeps me gnawing at a problem in physics when I might otherwise have knocked back my chair and decamped in frustration.

I am still a compulsive worker. Unless I am with Louise, or doing something for her, I am happiest sitting at my desk calculating. A few years ago, a journalist asked me how I managed to get so much done, and I told him that the trick was not going to church and not skiing. But I was just having fun with him. In fact, it is because, after all these years, I am still recoiling from my sophomore slump, always working close to Louise at home, and always with the TV turned on.

Recoiling from my sophomore slump, it became easy in my junior year at Cornell not to waste time. Physics and mathematics courses were becoming more and more sophisticated and engrossing. I took a graduate-level course in electrodynamics. The great thing about that course was not so much that I was learning about electric and magnetic fields, but that, at last, I was understanding relativity. Two years after Einstein had presented his paper, "A Special Theory of Relativity," the mathematician Hermann Minkowski, lecturing at Göttingen in 1907, described a way of writing the equations of physics that reveals the deep connection between space and time. Instead of locating an event with three coordinates (latitude, longitude, and altitude) to give its position in space, Minkowski argued that we should use four coordinates, the fourth being the time of the event. With this simple step taken, Einstein's results appear natural, almost obvious. And when we change the speed of travel of the platform from which we observe nature, the observer on a moving object does not experience the sensation of moving. We know the truth of this from our own experience. We do not feel the Earth's daily rotation, or its hurtling around the Sun.

And, just as space and time coordinates are helpfully combined in this way, so too are electric and magnetic fields. If Alice arranges that there is a uniform electric field in some direction and no magnetic field, and Bob observes the fields from a platform moving at right angles to the electric field, Bob will see an increased electric field and now also a magnetic field.

Minkowski's notation was used in the brand-new textbook for this graduate course, Arnold Sommerfeld's *Electrodynamics* (1952). This mathematical expression of the physics made special relativity much more clear to me.

That year, 1952–53, all the eager physics students at Cornell took a course on modern analysis given by the mathematician Mark Kac. We learned techniques of great power for solving mathematical problems, especially the sort of problem that arises in studying various kinds of waves. Every week, we were given a heavy load of problems to solve, which I thoroughly enjoyed. Yet, important as Kac's course was, it was notably old-fashioned. The text was Whittaker and Watson's *A Course of Modern Analysis*, for which, alas, "modern" meant "circa 1900." It had been first published in 1902. It contained nothing about group theory or topology, each of which was playing a huge role in the development of theoretical physics. These subjects apparently were not covered at Cornell at all. One graduate student, Harold McIntosh, took it upon himself to proselytize for matrix algebra, handing out mimeographed lecture notes on the subject to grateful undergraduates like me.

At the end of my junior year, Telluride awarded me "preferment" and invited me back for my last year at Cornell. Despite my amused misgivings about the Telluride sense of earnest mission, I had enjoyed living there, enjoyed the comic sense some of us Tellurideans shared with each other, and was glad to escape my lonely room in College Town.

The following summer, to be near Louise in New York, I managed to land a job at Bell Telephone Laboratories, the first time that anyone would pay me for work in a subject somewhat within the range of powers I was acquiring. I think Bell Labs hired me because I had taken a mathematics course on symbolic logic in my junior year.

Symbolic logic reduces the ordinary logical operations that we all use to a few precisely defined operations or propositions, statements that can be true or false. In the most elementary part of the course, we learned a formalism for logic known as Boolean algebra, introduced by George Boole in 1847. It happens that the hard wiring in telephone switching systems at that time reproduced the rules of Boolean algebra, with "on" and "off" replacing "true" and "false," as also occurs analogously in modern digital computers, with their ones and zeros.

It was a wonderful job. One of the problems I was assigned was to design a circuit that would count pulses in a convenient code that allowed automatic error detection. It led to my first paper, "Parallel Counter for Two-Out-of-Five Code." It was too unimportant to publish, but I like to suppose that it added one more to the millions of Bell Labs' patents.

That summer left me with a permanent sense of optimism about my future working life. Any sort of technical job I might have would be wonderful compared with the jobs I had held in earlier summers. Above all, at Bell Labs, when I had nothing better to do, I could go down to the library and read technical journals. Not that Bell Labs was a gentlemen's club. The expectation was that the visit to the library was useful. Yet it was the first professional job I had, within a company that respected the minds of its workers. I hoped some day to make a great contribution in physics, but even if that did not work out, I knew that I would always be better off than most of the workers of the world.

As for Bell Labs, that summer also left me with a high regard for it. At that time, it was one of the world's great research centers. At its headquarters at Murray Hill, New Jersey, were a brilliant cadre of physicists, including Philip Anderson, who revolutionized our understanding of the solid state of matter, and Claude Shannon, who began the application of mathematics to the flow of information. As an application of physics, the transistor was invented at Bell Labs by Bardeen, Brattain, and Shockley, and John Bardeen made a start in understanding superconductivity. At another location, in Holmdel,

New Jersey, in 1965, Arno Penzias and Robert Wilson were the first to detect a microwave radiation filling space, left over from the early universe.

I did not work at Murray Hill or Holmdel, but instead at Bell Labs' old building on West Street in Manhattan, the center of its switching systems division. I did not know it until recently, but it was in that building on West Street that in 1927 Clinton Davisson and Lester Germer had demonstrated the wave nature of the electron.

This kind of basic research was possible only because Bell Labs was supported by a regulated monopoly, AT&T, which could include research costs in the justification for its rates. It was a violation of free-market orthodoxy, to be sure, but it should have been vigilantly preserved. When a federal court in 1982 decreed the breakup of the AT&T monopoly, I feared that it would be the end of Bell Labs. And, alas, it was!

Having reached my senior year at Cornell, at last I learned quantum mechanics. The mystery of quantum physics had fascinated me since reading the books by Gamow and Jeans years earlier. Fortunately, our textbook was Leonard Schiff's *Quantum Mechanics* (1949), a clear and no-nonsense account of the theory. The book made plain how to use quantum mechanics to calculate properties of atoms and molecules and much else. I still have that copy of Schiff's book, and from time to time use it to look up a formula I might need.

With the confidence given to me by my newfound learning, I made an appointment to see Professor Hans Bethe, whose august presence at Cornell had attracted me there in the first place. I asked him if he could supervise me in a reading course on advanced quantum mechanics. Bethe gave me a kindly look but said that just then he was very busy. He certainly was. I did not know it then, but at that time, in the spring of 1954, Bethe was involved in the decision whether to develop a thermonuclear bomb. He was testifying in favor of J. Robert Oppenheimer in the secret hearings of the Atomic Energy Agency on Oppenheimer's security clearance.

I needed to go on to graduate school to study for a PhD in physics. I wanted to marry Louise, and I thought that, instead of heading for an American physics department, I should go to some European center for physics research, and enjoy a romantic honeymoon year abroad.

I applied for a Rhodes scholarship, but when I told the interviewer that I had no interest in athletics, he lost interest in me. (I consoled myself with the thought that no leader in physics has ever been a Rhodes Scholar, which as far as I know is still true.) I applied for a Fulbright scholarship, but was rejected for a mysterious reason I unearthed years later and will explain in a later part of this book.

Finally, I applied for a National Science Foundation (NSF) predoctoral fellowship. On this application, I had to indicate where I wanted to work. That year, a well-known theorist, Richard Dalitz, was visiting the Cornell physics department. When I asked his advice about physics in Europe, he suggested that I go to the Institute for Theoretical Physics in Copenhagen, familiarly called the Bohr Institute, named for its presiding genius, Niels Bohr. Of course I knew of Bohr as the first physicist who had figured out how to calculate the energies of atomic states, and I had heard of Copenhagen as a charming city and as the venue of great debates in the 1930s over the meaning of quantum mechanics. Dalitz also buttressed my impression that Copenhagen was just then becoming the headquarters of a new pan-European research organization, the Conseil Européen pour la Recherche Nucléaire (CERN). So that was my choice.

The NSF could not have cared less about athletics, and I was awarded a predoctoral fellowship. As far as I recall, the fellowship paid only $1,700/year, and $250 more for fellows already married, but the dollar was very strong then against European currencies, and it would be enough.

Louise and I both graduated with "distinction in all subjects," Cornell's equivalent of summa. (Louise had been inducted into Phi Beta Kappa in her sophomore year. Only late in life was I made an "honorary" member.) Shortly before graduation, I was in a poker game at Telluride when one of the players, mathematics professor Mark Kac, told me that my team had won first place in a national mathematics contest, the William Lowell Putnam Competition. As an individual, I was awarded only an honorary mention, but received a beautiful little gold medal, my first medal, that Louise sometimes wears on a necklace.

4 Copenhagen

For complicated reasons having to do with my need for a married fellow National Science Foundation (NSF) grant, which was more substantial than the grant for a bachelor fellow, we were nominally married at a private ceremony by a judicial friend of my father's. Louise was booked for a tour of Europe with her family, but would rejoin me in London, where we would be properly married.

I arrived in Copenhagen and reported my existence to the Bohr Institute. I went the next day to pay my respects to Niels Bohr. He was then almost seventy. He had made significant advances in physics. In 1913, at Manchester, he was the first to explain that the bright and dark lines that Fraunhofer had discovered in the spectra of stars and flames are due to the emission and absorption of particles of light in transitions between atomic states. He was the first to use quantum theory to calculate the wavelengths of these spectral lines, or to see them as connected with atoms. In the 1920s and 1930s, he presided over debates with Einstein and others over the meaning of quantum mechanics. After Denmark was invaded by Germany in 1940, he had had a historic visit from Heisenberg, dramatized in Michael Frayn's play, *Copenhagen*. He then escaped to Britain, and then to America, where he helped with the Manhattan Project at Los Alamos, the development of the atomic bomb.

David Frisch was one of the American physicists there. Frisch was visiting from MIT. He alarmed me when he asked me a question that was to change the course of my life. He asked, "What sort of research are you doing, Steve?"

I protested that I was a first-year graduate student, and was not ready to start research. I had to take my graduate courses. I had to spend this year in Copenhagen reading physics books and articles.

I was under the illusion that, in order to be able to do anything new, I had first to know every relevant thing that had already been done. Dave quickly disabused me. No, he said, this was a research institute, and everyone had to be doing research. Dave proposed that I work on nuclear alpha decay. Alpha decay is one of the two kinds of radioactivity (the other, unsurprisingly, is called beta decay) that were identified by Rutherford at McGill University in the 1890s. As Rutherford showed, alpha decay is the spontaneous emission of helium nuclei by unstable nuclei, such as uranium 238 and radium 226. At that time, I had barely heard of alpha decay, so I did not understand what the problem was on which I was supposed to work. I read every research article I could find at the Institute library. Oddly enough, to this day, I have not been able to get anywhere with Dave's proposed investigation. But even though I made no progress on alpha decay, I had the experience of reading research articles in an area that was previously unknown to me. I gained confidence that I could pick up what I needed in doing research without mastering the whole subject in advance.

In the late 1940s, Feynman, Schwinger, and Tomonaga had made a great advance in understanding quantum electrodynamics, the quantum field theory of radiation and electrons. That was just the sort of new thing on which I wanted to work. So, in the callowness of youth, I had therefore brought with me the new edition of Heitler's *Quantum Theory of Radiation*. I soon found that, as in earlier editions, Heitler was describing the state of the subject as he knew it in the 1930s. There was nothing in it about the new advances.

Then I heard that at the Institute there was a young Swedish theorist, Gunnar Källén, who had proved an important theorem about quantum electrodynamics. Källén showed that some of the constants in the equations of the theory had to be infinite. Oppenheimer and Waller had independently found, in the early 1930s, that some calculations in quantum electrodynamics give infinite results for observable quantities like energies and probabilities that, of course, must be finite. That became the bone in the throat of theoretical physicists for over a decade.

FIGURE 4.1 The author as a young researcher

The breakthrough in the late 1940s by Feynman, Schwinger, and Tomonaga was to show that all infinities in observable quantities would be canceled if one supposed that some of the constants in the equations of the theory were infinite. For instance, one of these infinite constants is what is called the "bare mass" of the electron, the quantity "m" that we put in the equations to stand for what would be the electron mass if the electron were not surrounded by a cloud of photons (particles of light) that it was continually emitting and reabsorbing. But the electron is surrounded by such a cloud, which makes its own infinite contribution to the electron mass that we observe. The infinity in the bare mass cancels the infinity associated with this cloud, and the observed mass is perfectly finite. Using this technique of canceling infinities, known as renormalization, theorists were at last able to do calculations of effects in atomic physics. And these were beautifully confirmed by experiment.

Despite this success, it is distasteful to suppose that the equations of physics involve constants of nature that are infinite. It had been hoped that these infinities are an artifact of the kind of approximations that are always made in practice, and somehow could be made to go away. We needed to find a way we could dispense with these approximations in our calculations. But Källén's work seemed to close off this possibility.

I introduced myself to Källén, and asked if he would be interested in suggesting a research problem on which I could work. Indeed, he would. Källén told me that there was a young theorist at Columbia University, Tsung-Dao Lee, who had claimed that infinities could be banished by allowing a certain departure from the usual rules of quantum mechanics, known as an indefinite metric. Briefly, the idea was to allow intermediate states with negative probabilities in physical processes, though of course the probability of getting any specific final state must always be positive. Lee had presented a model theory that seemed to bear this out.

Källén was not going to put up with anyone getting rid of infinities so easily. I think that for him this was like a fundamentalist preacher being told that there is no hell. Earlier that year, Källén, with his mentor, Wolfgang Pauli, had shown that the Lee model did in fact predict negative probabilities for physical processes if the constant in the theory that governs the strength of forces was sufficiently large. Källén suggested that I should study the Lee model and see what else was wrong with it.

So now I had a research problem! In some respects, it was a good assignment for me. I understood what the problem was, and it was within my capabilities. Probabilities and energies could be calculated in the Lee model without having to make any inventive approximations. But working on this topic did not help to bring me up to speed with the recent advances in quantum electrodynamics. The reason that the Lee model could be solved exactly was that only a limited number of particles would appear in the model, in the intermediate states of any process. A limit on the number of particles in Lee's model was not in accord with Einstein's special theory of relativity. Relativistic theories like quantum electrodynamics, in principle, do not allow any limit on the number of particles in intermediate states. This is what makes quantum electrodynamics impossible to use without approximations, and until the new work of the late 1940s impossible to use even with approximations for any but the simplest problems.

I asked Källén to suggest anything I might read about the new techniques for calculation in quantum electrodynamics, specifically the method known as Feynman diagrams. He told me not to bother; Feynman diagrams were unnecessary. He was the wrong one to ask about this. Källén worked in a highly mathematical tradition, in which the important thing is to prove theorems, not to do approximation calculations to get results that might be compared with experiment.

Even though Källén was not an ideal mentor, he was very kind. He made a point of introducing me to Wolfgang Pauli when the great man visited Copenhagen. Pauli had been the first to suggest the existence of the neutrino, in order to make sense of experimental results on beta decay, and he had been the author of the exclusion principle, according to which no two electrons could have the same momentum and spin. (It is the exclusion principle that provides an explanation for the periodic table of the elements.) Not only was Pauli capable of such leaps of imagination; he was mathematically stronger than his great contemporaries. The earliest version of quantum mechanics was a formalism known as matrix mechanics. It had been invented by Heisenberg, but Heisenberg had been unable to use it to do any but the simplest calculations. It was Pauli who showed how to use matrix mechanics to find the energy levels of the hydrogen atom, the crucial problem for testing the new theory.

Pauli was also famously cocky. There was a story about him that went around at the Bohr Institute, that once when Pauli was a student at Göttingen he had attended a physics lecture by Einstein. No one after Einstein's talk wanted to be the first to comment, until Pauli spoke: "What the learned professor has said is by no means so stupid" ("gar kein so dumm"). Once, when presented with an incoherent physics article, Pauli commented that it was so bad that it was "not even wrong."

Källén and his wife invited me to bring Louise to dinner at their house. Of that evening, my most vivid memory is finding in the bathroom hand towels embroidered with a famous formula, known as the Dirac equation. When I asked about this, Källén told me that the towels were a gift from Pauli.

Near the end of the autumn term, Louise and I were invited to a dinner that Niels Bohr and his wife Margarethe were giving for all the faculty and visitors at the Institute. The dinner was in the conservatory, the huge glass hothouse of the grand house that had been provided to Bohr by the Carlsberg Foundation, funded by the makers of Denmark's popular beer. Perhaps as a display of noblesse oblige, or because she was so pretty, Bohr placed Louise next to him at the dinner table.

Bohr had an ironic wit. There is a famous story: When Einstein criticized the introduction of probability into the equations of quantum mechanics by objecting that God does not play dice with the cosmos, Bohr answered that Einstein should stop telling the almighty how to behave.

A less well-known story went around the Institute the year I was there. A visitor to Bohr's country house noticed an iron horseshoe nailed above the door, and exclaimed, "But Professor Bohr, you don't believe that horse shoes bring good luck, do you?" Bohr answered, "No, of course I don't believe it. But they say it works even if you don't believe in it."

I planned to obtain my PhD from Princeton. I booked cheap passage for us on a slow Danish freighter. I had plenty of time to finish my work on the Lee model, and found that indeed there was something else wrong with it: The energy of interaction between two of the particles of the theory turned out not to be a real number, but to involve "i," the square root of minus 1, which makes no sense.

5   Princeton

I took a summer job in the Atomic Beam Group in the Princeton physics department. Atomic beams are used to study atoms in isolation as they drift through a near vacuum, free from the perturbations of other atoms that would be nearby in a liquid or solid. These beams of atoms have to be focused by traveling between the poles of magnets of various sorts along their paths. In different experiments, the magnets have to be placed in different sequences along the beam path, and it is necessary for experimenters to calculate the trajectories of atoms for each sequence. This was my job.

In their previous work, the members of the Beam Group had calculated the trajectories of beam particles separately for each sequence of magnets. I realized that each magnet could be characterized once and for all by a 2×2 matrix, a square array of four real numbers. Knowing this matrix, it was simple algebra to calculate, from the displacement and rate of change of displacement of the particle as it entered the magnet, the displacement of a beam particle from the center of the beam as it left the magnet, and the rate at which that displacement was changing. You could carry out a similar calculation for a whole sequence of magnets using a 2×2 matrix that is found by multiplying the matrices for each magnet in the same order as in the sequence of magnets, following the standard rules for multiplying matrices.

I was pretty proud of this idea, and thought of writing it up for some technical journal. Then one day in the Princeton University bookshop, I noticed a book with the title (as well as I can remember) *Matrix Methods in the Design of Optical Instruments*. Without even opening the book, I knew that, though the book would deal with beams of light instead of beams of atoms and with lenses instead of magnets, it must describe the same method I had developed. So all I had done was to deploy a known method.

My first year at Princeton went very well. The personal relations of graduate students in physics with the younger faculty were closer at Princeton than I have since seen anywhere else. We usually lunched together, and shared ideas about the latest physics news. In November, I sent in my paper on the Lee model for publication in the *Physical Review*. With my year at the Bohr Institute behind me and a research paper in press, I felt more like a colleague than a student. I did not hide this feeling, and spoke up frequently in class and in research seminars. Of course I was insufferable, but I enjoyed feeling like a grown-up.

Pauli made a visit to Princeton that year. There were no other graduate students and few faculty members who had met him, so I was given the job of taking him around Princeton. Pauli was old and not outgoing, but I felt honored just to be in his company.

The Princeton physics department, though offering formal traditional courses for its graduate students, was untraditional and outstanding in its emphasis, for its graduate students, on research rather than course work. It was a situation for which my stint at the Bohr Institute had fortuitously prepared me well. On the other hand, as in every place of learning, formal courses will be of mixed quality, reflecting the personality and point of view of individual professors.

Eugene Wigner was the most distinguished member of the department's faculty. He had grown up in Budapest, was educated in Germany, and then had come to the US in the early 1930s. In his fifties when I first knew him, he was small, bald, formally dressed, and excessively polite. Once, a few years later, at a conference, when a group photo was being taken, I watched as he vigorously demurred from being placed in the front row. Somehow, however, when the photo was finally taken, there he was, front row center.

Wigner had done more than anyone since Einstein to bring an understanding of symmetry principles into physics. When years later I wrote my treatise, "The Quantum Theory of Fields," its treatments of spinning particles and of resonance would be based on Wigner's

FIGURE 5.1 Eugene Wigner, c.1956

work. Yet his class on nuclear physics was perhaps the worst I have ever taken. Of course Wigner was a master of the subject, but he kept worrying about whether his lectures were being understood by students. He kept peering at us to see if we were following what he was saying, and the course was glacially slow. Somehow he had got the impression that I did understand him, and I received a very great compliment: When Wigner had to be away from Princeton for a week, he asked me to take over his class.

I took a course on general relativity from Valentine Bargmann. Here too I had reservations, but about point of view rather than pace. Bargmann had escaped to America from Germany in 1933, and worked with Einstein. Like many other theorists of his generation, he regarded general relativity as the application to gravity of the mathematical theory of curved space-time. So in studying the motion of bodies in a gravitational field, it was taken for granted that they would travel on geodesics, the curves that generalize the straight lines of ordinary Euclidean space. I did not doubt the validity of general relativity as taught by Bargmann, but I thought it needed to be based on something more physical than analogies with geometry. As I will explain in a later chapter, eventually I worked this out for myself when I began teaching general relativity at Berkeley.

My Princeton professors were casual about grading. Arthur Wightman was a starring figure on the mathematical side of quantum field theory and, in my first year at Princeton, gave the course on advanced quantum mechanics. For some reason, he had recently become interested in the statistical properties of polarized light, and lectured at length about Stokes parameters, the numerical quantities used to describe the probabilities of various polarizations. This had nothing to do with the exciting recent advances in quantum electrodynamics. He made up for this with his final exam. It was a take-home problem, to reproduce Julian Schwinger's classic calculation in quantum electrodynamics of the strength of the electron's magnetic field. After we had handed in our calculations, Wightman told us that he was going to go home and burn them without reading them. The point was to give us the experience of doing this sort of calculation, not to judge us. I don't recall if I was happy about this at the time, but this exercise turned out to be the most valuable part of all my course work as a graduate student.

The following year, the course on advanced quantum mechanics was given by John Archibald Wheeler. Wheeler was a courtly figure, politically conservative and scientifically imaginative. As a young man, he had worked with Bohr to develop a theory of nuclear fission, and had then been active in the Manhattan Project. After the war, his interests turned to general relativity, especially the theory of black holes. He inspired a generation of younger relativists. Wheeler and I became friends in the 1980s, when we were both on the faculty in Austin, though we continued to disagree about arms control, manned space flight, and cosmology. I could not go along with his views on black holes, but it turned out that Wheeler was certainly right about black holes. My skepticism was to be blown away by explosive developments in the twenty-first century.

The course that Wheeler gave on advanced quantum mechanics was very much in the spirit of his most famous student, Richard Feynman. It emphasized an approach to quantum mechanics invented by Feynman, known as the path integral formalism. In this approach, to find the probability of a transition from one state to another, as in a scattering process, you have to do a sum over all paths that the particles of the system can travel between their initial and final positions. In the cases that Feynman had studied, the results agreed with those of ordinary quantum mechanics, but it was a dramatically different and, to some extent, convincing way of doing calculations. Feynman had developed this approach as an alternative to ordinary quantum mechanics, on the basis of his sheer physical intuition, and that is the way Wheeler taught it.

I was repelled by this intuitiveness. It seemed to me that there is very little that is authoritative in anyone's intuition. I was half-right to be skeptical. As I would learn over a decade later, from the work of Freeman Dyson, the path integral formalism can be derived as a consequence of quantum mechanics, without relying on anyone's intuition. Indeed, in the simple version that Feynman had invented and Wheeler taught, it does not always work. To be sure, the path integral formalism does make possible certain calculations in quantum field theory that can't be done any other way. But none of that came into Wheeler's course.

I took away from this course a distaste for the path integral formalism. That was another great mistake. As acknowledged in a later chapter, my dislike of the path integral formalism would stand in my way at an important point in my work.

I also took a course on modern algebra from a renowned mathematician, Emil Artin. It provided a classic example of a recurring problem in communication between mathematicians and physicists. Mathematicians often adopt a lapidary style, in which theorems are presented with no comments on the side that are not needed in the proof. They admire the style of the great Carl Friedrich Gauss, who in his articles hid any hint of the motivation behind his brilliant proofs. Artin was like that. He lectured about something called an "exact sequence" without bothering to explain that this provided a method that is useful in topology. I dropped the course.

I buckled down and finally began to learn modern quantum electrodynamics. I saw the necessity of digging into the literature on my own, and even then I didn't entirely trust the authorities I was reading. Feynman seemed to me too hand-waving, airily taking flight above difficulties. Concededly, that sort of thing can be a marvelous aid to progress. But it can be taken too far. On the other hand, Schwinger and Tomonaga seemed to me unnecessarily formalistic and plodding. I finally found enlightenment in the papers published by Freeman Dyson in 1949. Dyson was one of the great minds in physics at the time, and also happened to be one of the best writers in the field. It was Dyson who had re-derived Feynman's rules for calculating the rates of various processes, working directly from the principles of electrodynamics and quantum mechanics, with no need for path integrals.

More than that, Dyson had given at least a plausible argument that the trick for getting rid of infinities known as renormalization would work no matter how complicated the emissions and absorptions of particles in the process. He was on the path toward making renormalization legitimate, not merely necessary.

I did not doubt what Dyson was saying, but I thought a more thorough argument was needed. I spent much time studying papers on the subject by the Pakistani physicist Abdus Salam. Although Salam seemed to me to have closed some loopholes in Dyson's argument, I did not find even his work entirely convincing.

I went over to the Institute for Advanced Study, where Dyson was a professor, to ask him about renormalization and the infinities. Despite my admiration for his work and his writing, he seemed to me brusque and unhelpful.

Dyson was a donnish Englishman who had worked for Bomber Command in the war, and had strong views about everything, from strategic bombing to quantum field theory. He wrote as beautifully for the general public as he did in physics. He published a brilliant meditation on Bomber Command in the *New York Review of Books*. Our interactions later would be much friendlier, and in one case he would be extraordinarily helpful to me. But he always seemed to me perverse in the pleasure he took in pooh-poohing whatever was becoming established. He seemed determined to oppose any consensus that he sensed forming among scientists. He was against the nuclear test ban treaty and a skeptic about climate change. He seemed to me to be like the physicist-philosopher Ernst Mach, who in the early years of the twentieth century complained that other scientists had adopted a view of the reality of atoms as a sort of dogma, even though atoms were little else but an idea, a mental construct. Both Dyson and Mach were brave and eloquent and willfully wrongheaded.

At that time, the director of the Institute for Advanced Study was Robert Oppenheimer. In the 1930s, Oppenheimer was the first American theorist to make significant contributions to the new discipline of quantum mechanics, and then, in the war, he was the successful director of the Los Alamos National Laboratory. He was an inspiring figure to my generation of physicists. At Princeton, I had a chance to observe Oppenheimer in action. I described this in a review of Jeremy Bernstein's biography of Oppenheimer:

> When I was a graduate student at Princeton I used to go over to Building E of the Institute to attend physics seminars. Oppenheimer always sat in the front row, asking questions that demonstrated that he knew as much about the speaker's subject as the speaker. Of course he was showing off, but no one else could have gotten away with it. He did know as much as the speaker.

That autumn of 1956, I began to spend time talking about physics with a pair of theorists who shared an office in the crowded Palmer Physical Laboratory, Sam Treiman and Marvin (universally known as "Murph") Goldberger. Treiman had received his PhD in physics at Chicago just four years earlier, but he knew a great deal about the weak interactions of elementary particles, such as the radioactive process of beta decay, most of which was new to me. Murph was a more senior figure. He was a leader in deriving exact results known as dispersion relations, governing the otherwise murky strong interactions among nuclear particles and mesons.

Sam and his wife Joanna invited Louise and me home to dinner. We had a dinner invitation also from Carl Levinson and his wife, whom we had known in Copenhagen. Kindness like this from faculty to a graduate student is unusual in physics, and may well have been unusual even at Princeton. I felt very grateful and rather flattered.

I liked doing physics in Princeton, but Louise and I were increasingly dissatisfied with the lack of urban life of the sort we had enjoyed in Copenhagen. Of Princeton, Sidney Coleman had once remarked that it was wonderful living there: For entertainment, you could go down to the railway station and wait for a train to come in. One evening when we were dining out in one of the two Italian restaurants, Louise broke it to me that she was too bored in Princeton (she had been involved with a theater company and had also worked in the psychology department, a job consisting in cheering up the peculiar old men in Princeton's psychology department). She was leaving Princeton and would be visiting with her mother until I got my PhD and a job anywhere else. So now I was alone in Princeton. I set out to get what I intended would be the fastest PhD degree in the university's history.

My article on the Lee model was too unimportant to serve as a PhD thesis, and my effort to clean up the justification for renormalization theory had not yet gotten very far. Searching for a research problem, I browsed through a recent book on elementary particles. Reading about the current theory of weak nuclear interactions, it struck me that an important matter was being left out.

Consider the paradigmatic process of nuclear beta decay, in which a neutron spontaneously decays into a proton, an electron, and a neutrino. Neutrons and protons interact strongly with lighter particles called pions. So while a neutron is turning into a proton, it can emit and reabsorb any number of pions. This is similar to the emission and reabsorption of photons by an electron. As seen in Schwinger's calculation (and Wightman's homework problem), this process modifies the magnetic field of the electron.

The strength of the electromagnetic processes involved in emitting and absorbing photons is characterized by a fairly small number, known as the fine structure constant. It is equal to about 1/137. The effect of emitting and absorbing a single photon is suppressed by a factor of 1/137; the effect of emitting and absorbing two photons is suppressed by a factor of $1/(137)^2$; and so on. As more and more photons are included, the calculations get more and more complicated; but the effects get smaller and smaller, so it is possible to get very accurate results by just taking into account only a small number of photons. In Schwinger's original calculation, there is only one.

However, strong interactions are strong. The quantity that characterizes their strength is not small, like 1/137, but more like 1. The effect of emitting and absorbing a large number of pions in beta decay is proportional to a large power of 1, which of course is still just 1. There was no way to make approximations, and thus no way to calculate anything.

The equations that physicists were using in analyzing processes like nuclear beta decay did not explicitly take account of the emission and reabsorption of unlimited numbers of pions, but they worked pretty well. I concluded that these equations could not be the fundamental equations governing the fields of particles like protons, neutrons, electrons, and neutrinos, but had to be including, implicitly, the effects of strong interactions with pions.

This may just seem like a matter of words, but this perspective led to a substantive new conclusion. When theorists introduce a field theory of weak interactions such as Enrico Fermi's 1934 theory of beta decay, they are not really free to limit its content on grounds of beauty or simplicity, as if it were a fundamental law of nature. Since these theories, to be useful, must implicitly take into account the effect of strong interactions, any complication that can be produced by processes like pion emission and reabsorption and that is not forbidden by conservation laws must be included. It is like the law governing the society of ants in T. H. White's *The Once and Future King*: EVERYTHING THAT IS NOT FORBIDDEN IS COMPULSORY.

In particular, Fermi's field equations had to be supplemented with a new term, called an induced pseudoscalar, produced for instance when a proton turns into a neutron with the emission of a pion, which then undergoes decay into an electron and a neutrino. This makes little difference in nuclear beta decay, but it can be important in a process like the conversion of a proton into a neutron by the absorption of a sort of heavy electron called a muon.

At the time, I did not think much of this work, but looking back I can see that it was an anticipation of what I would later call effective field theory. In retrospect, this gives me a warm feeling about it, which I definitely did not feel at Princeton.

Anyway, I thought that this work might serve as a PhD thesis. I had been spending a good deal of time chatting about weak interactions with Sam Treiman. One day around the beginning of 1957, I bravely put it to him that, if he did not want to be my thesis advisor, he had better tell me right away. When Treiman accepted me, I wrote up my ideas as a thesis and as an article titled "Role of Strong Interactions in Decay Processes" that I sent to the *Physical Review* that March.

I had not been Treiman's first PhD student. My friend Nick Khuri had started earlier with him. But I could not bear living without Louise. I was Treiman's first student to finish work for a PhD.

FIGURE 5.2 Sam Treiman

Treiman had become friendly with T. D. Lee, in Chicago. In 1957, Lee was the senior theorist at Columbia. Treiman quietly arranged with Lee that I would be an instructor there starting at Columbia in the fall of 1957 (it would not be possible to manage things this informally today). I sold our furniture and our car. I took my final oral exam, and left Princeton for New York and my girl that spring.

6 Manhattan

In the spring of 1957, we moved to New York and my job at Columbia University. Columbia was then at the center of the most exciting progress in physics.

T. D. Lee was the senior theorist at Columbia. He showed no signs of resentment that I had found things that were wrong with the Lee model. Probably he had not noticed, for he had more important things on his mind.

In 1956, Lee, with C. N. Yang, had made the remarkable conjecture that the laws of nature did not always respect the symmetry between right and left. When as a graduate student I heard about this idea, I thought it was absurd. Of course, the history of life did produce differences between left and right – our hearts are on the left, and we can only digest sugars that rotate the plane of polarization of light to the right. But these are historical accidents. It had always seemed to me and to almost all physicists that the laws governing physics and chemistry, the laws underlying everything, did not distinguish left from right. As an undergraduate at Cornell, I had learned how physicists use this principle to figure out what sorts of transitions between atomic states were possible, and the results were always borne out by experiment. But Lee and Yang recognized that all the evidence for the symmetry between left and right had come from processes involving electromagnetic or strong nuclear forces; there was no evidence that the symmetry is respected by the weak forces, the forces responsible, for instance, for nuclear beta decay. There was one puzzling feature of the weak decay of particles called K mesons that could be interpreted as a sign that the left–right symmetry is not respected by weak forces, but it could be explained in other ways as well. Lee and Yang went beyond speculation and proposed experiments to test their idea. In

FIGURE 6.1 T. D. Lee in 1957

1957, just as I was arriving at Columbia, experiments were showing that Lee and Yang were right.

Most of this experimental work was done by people at Columbia, which was not surprising since Columbia at the time had the greatest assemblage of experimental talent in the world. A group headed by Chien-Shung Wu found that the pattern of beta decay of cobalt-60 depends on whether the nucleus as viewed from the emitted electron is spinning to the left or right. This was experimental physics in the old style, done on laboratory tables, using magnetic fields to choose the direction of spin of the cobalt nucleus.

At about the same time, Richard Garwin, Leon Lederman, and Marcel Weinrich found a left–right asymmetry in existing records from accelerator experiments of the tracks left by charged particles in the chain of decays of pions to muons to electrons. (The pion decay work was also done independently at Chicago by Jerome Friedman and Valentine Telegdi.) Also at Columbia were Mel Schwartz, who had preceded me by a few years at Bronx Science, Jack Steinberger, who had escaped from Germany in 1934, and Carlo Rubbia, who arrived at Columbia a little after my arrival.

Most of these experimenters became friends of mine, and most became famous in elementary particle physics. The reader, if not a physicist, may not recognize these names, but to a physicist, it is as if I were describing a movie with a cast including Clark Gable, Humphrey Bogart, Cary Grant, John Wayne, and Spencer Tracy. (I omit female superstars reluctantly, but few women were involved in the physics I have been talking about, with Chieng-Shung Wu being one such exception.)

The most famous physicist at Columbia was I. I. Rabi. He had discovered nuclear magnetic resonance, the basis for MRI scans. During the war, Rabi had played a leading role in developing microwave radar at MIT's Radiation Laboratory. His group had made the measurement of the strength of the electron's magnetic field that had then been explained in a calculation by his student Julian Schwinger.

FIGURE 6.2 John Archibald Wheeler, I. I. Rabi, and Eugene Wigner

Rabi made a famous remark about an elementary particle, the muon. This particle was discovered in cosmic rays in 1937, and at first was thought to be the "meson" that Hideki Yukawa had recently proposed as the carrier of the strong nuclear force. But after the war, it was realized that these particles do not participate at all in the strong nuclear interactions. They behave just like electrons, but are about 210 times heavier. No one knows why they exist. About the muon, Rabi had asked, "Who ordered that?" Rabi was quoting a recurring remark of Columbia physicists. Once a week, some of us would amble over to upper Broadway for lunch at one of the very good Chinese restaurants there. As is often the custom in Chinese restaurants, each of us would order one dish, which would sit on a turntable at the table's center for all to share. Some of these dishes, like sea cucumber, were pretty strange looking. Hardly a week would go by when someone did not point at a dish and ask, "Who ordered that?" So it happened that, when a particularly unlovely equation or phenomenon would appear in our work, we found ourselves saying, "Who ordered that?"

At Columbia, I continued my effort to show that the redefinition (or "renormalization") of constants like the charge and mass of the electron really did remove the infinities encountered in any calculation in theories like quantum electrodynamics. But in the Columbia environment, I felt I had to extend my range and work also on problems in the theory of elementary particles, and in particular on their weak interactions. I have always liked to work in areas of interest to colleagues at the place where I am working.

I wrote papers on symmetry principles and weak interactions, some with Sam Treiman, mostly not very important. In one paper with my old friend Gary Feinberg, who was then at Brookhaven National Laboratory, and a new friend, Pasha Kabir, I addressed the question of why the muon did not decay into an electron, emitting a photon. We showed that one mechanism that appeared to produce such a decay in fact did not. The conclusion was correct, but it was the reasoning that turned out to be important years later in my work on the Standard Model of elementary particles. It did not attract much attention at the time.

Gary Feinberg wrote a much more timely paper, examining the question of muon decay into electrons under the popular assumption that the weak interactions are carried by a hypothetical charged particle that was too heavy to have been observed but had even been given a name, the *W* particle. (*W* stands for "weak," not Weinberg.) The muon was known to decay slowly into electrons with the emission of a pair of neutrinos. Gary showed that if this occurred by the muon emitting a *W* particle and a neutrino, followed by the *W* particle turning into an electron and another neutrino, then instead of the two neutrinos emerging in the decay, they could annihilate each other, with the energy in the muon's mass carried away by a photon. Since this was not observed, Gary's paper raised a serious problem for weak interaction theory. (Later it was shown by Lederman, Schwartz, and Steinberger that the two neutrinos emitted in muon decay belong to different species and so cannot annihilate each other, which was the solution to the problem raised by Gary.)

In the autumn of 1958, Gary and I attended a conference on weak interactions at Gatlinburg, Tennessee. I was excited because it was to

FIGURE 6.3 Gary Feinberg

be the first national physics conference I would attend, and also because one of the speakers would be Murray Gell-Mann. Murray was only a few years older than me, but had already earned an international reputation. Murray was legendary for his celebrated physics papers, some with Richard Feynman.

I had been spending some time studying one of Murray's papers considering the effect of strong interactions on weak interactions like beta decay. I had found a contribution to the interaction among particles in beta decay that I had studied in my thesis. This contribution was the "induced pseudoscalar," produced by the strong interactions among protons and neutrons. Gell-Mann found another effect that he called "weak magnetism." What was remarkable to me about Gell-Mann's paper was not so much this finding, but the depth of the understanding of nuclear physics that he showed in analyzing its experimental implications. I had a lot to learn.

In the course of studying this paper, I realized that Murray was implicitly making an assumption about the symmetry properties of the weak interactions. If this assumption were not made, there would be two other terms in the beta decay interaction, which I called second-class currents. When Murray spoke at Gatlinburg, I could see that, though he did not say so, he was still making this symmetry assumption and was ignoring the possibility of second-class currents. In the discussion session after his talk, I pointed this out, acknowledging that Murray might well be right but that it was necessary to recognize that a symmetry principle was being assumed and might not be valid. Murray thought for a little while. Time passed. At length, Murray answered simply: "No." The physicists present were silent, probably embarrassed by my temerity. More time passed. Then I replied, also simply: "Yes." The audience exploded in laughter. Murray changed the subject. I believe he never forgave me.

Some experimentalists took the possibility of second-class currents seriously enough to look for them. Murray was right – they do not exist. But I was also right, that to explain this it was necessary to make a special assumption about the symmetry properties of the weak interactions. The rationale for this assumption was finally understood with the development over a decade later of the Standard Model of elementary particles.

This episode began my long and rocky interaction with Murray. He was a great figure in physics, who in the 1960s recognized a symmetry of the strong interactions he called "the eightfold way." (This symmetry was also identified independently by the Israeli physicist Yuval Ne'eman.) Murray then explained it in 1964 with his theory of quarks (proposed independently at Cal Tech by George Zweig). Murray was intensely competitive both with more senior figures like his Cal Tech colleague Feynman and also with more junior theorists like me.

The last time that I met Murray was a few years ago, in Austin. Murray, an enthusiastic birder, had come to look at birds. There is hardly any time in the year when flocks of birds are not migrating north or south through Austin. On this visit to Austin, I took Murray to lunch. For some reason, at lunch, he made some disparaging remarks about Freeman Dyson. Murray indeed had made even greater contributions to physics than Dyson, but as a graduate student I had been struck with admiration for Dyson's work on quantum electrodynamics and had learned much from it, so I strongly disagreed. Uncharacteristically, Murray backed off, acknowledging that Dyson had done some good things. Murray had mellowed.

At Gatlinburg, I had some conversations with Robert Marshak, the senior theorist at the University of Rochester. A little earlier, with George Sudarshan, Marshak had made an important suggestion about the form of the beta decay interaction, which Feynman and Gell-Mann made independently at about the same time. I felt flattered when Marshak invited me to come to Rochester and work for a while with him and Sudarshan on weak interactions. A few weeks later, we took a Capitol Airlines flight to Rochester on a Vickers Viscount. (Reader, do not worry, I am not going to mention the airline and aircraft for every trip in my working life. But this was the first time that anyone had ever invited me to fly somewhere at their expense, so it is vivid in my memory.) We wrote a couple of papers about weak interactions, nothing very important, but I learned a lot.

At Columbia, I shared an office with Stanley Mandelstam, from South Africa, and Alberto Sirlin, from Argentina. We became good friends.

At the Chinese lunches at Columbia, I came to know the nuclear theorist Robert Serber. Bob had collaborated with Oppenheimer at Berkeley and Cal Tech before the war and had worked in the war on the Manhattan Project. His friendly wife Charlotte was full of stories about their days on the mesa at Los Alamos. Oppenheimer had made her chief librarian there. She was vivacious and had a wonderful sense of comic timing. She told us how a group from Los Alamos would go out to the La Fonda bar in Santa Fe in order to mislead the natives about the work at the laboratory. They would converse loudly about electric submarines. Unfortunately, that story had no effect. Everybody seemed to know what was going on at Los Alamos.

Bob Serber had been on the first American team to visit Hiroshima and Nagasaki after the Japanese surrender. He told a revealing story about Oppenheimer. Shortly after the war, he and Oppenheimer and a number of leading physicists were together at a conference in Seattle, and they made an excursion on a vessel on Puget Sound. When fog settled in and navigation became impossible, someone asked what would be the effect if the boat foundered and a whole generation of physicists was lost. Oppenheimer answered, "It would do no permanent good."

I became friends also with a Columbia professor, Norman Kroll, who had done some very important calculations in quantum electrodynamics. Norman won my heart when he said that he adopted the point of view about the role of strong interactions in beta decay that I had advocated in my PhD thesis. A few years later, I was surprised when Norman told me that he was going to accept a professorship at the new La Jolla campus of the University of California, because he liked going to the beach. I could not imagine anything better than being able to stay in Manhattan as a professor at Columbia.

It became common knowledge in late 1958 that Columbia was going to have just one tenure-track assistant professorship available in theoretical physics. It was first offered to Mandelstam. This was no surprise, for Stanley had made a big splash with a formula for the scattering due to strong interactions known as the "Mandelstam representation." At that time, physicists had pretty well given up on the hope of calculating strong interaction processes using the kind of approximation that had worked so well in quantum electrodynamics. Instead, it was becoming popular to write formulas for the rates of strong interaction processes that embodied general principles including principles of symmetry like those of special relativity and the principle that the probabilities for each possible outcome of a scattering process always had to add up to 100 percent. These formulas involved unknown quantities, but it seemed at least a good start. Murph Goldberger at Princeton had derived such a formula, known as a dispersion relation, for the dependence of pion–nucleon scattering on a single variable, the energy. Then Mandelstam had extended it to two variables, the energy and the momentum transferred between the particles. At the time, the vital center for this sort of research was a group headed by Geoffrey Chew at the University of California at Berkeley. They offered Mandelstam a professorship, and he accepted.

This left three contenders for the position at Columbia: my old friend Gary Feinberg, Alberto Sirlin, and me. One day in the spring of 1959, I received a phone call from Gary. He told me that he had been offered the assistant professorship, and he had accepted. I was very disappointed, but at the same time genuinely glad for Gary. He had wanted so much to be in New York that he had not left the city to go to college, as Shelly Glashow and I had done. Also, Gary deserved the offer. His paper on the decay of muons into electrons was more important than anything I had yet done.

The department chairman, Henry Foley, told me that I would be welcome to stay on as a postdoc for at least another year, but he gave me no hint that there would be another opening for an assistant professorship. Staying on would have been the easiest choice, and I was prepared to stay, but Louise talked me out of it. I recall her concise statement of the situation: "We have to leave in a huff."

I had offers of assistant professorships from several physics departments. But they did not seem to me to be in the front line of current research, not like Columbia, Princeton, or Berkeley. I also had an offer of a nonacademic long-term position in the Radiation Laboratory at Berkeley. Murph Goldberger assured me that this would lead to a professorship on the Berkeley campus. Swayed in part by this seemingly inside information, I accepted the Rad Lab's offer. Murph's prediction was to come true, but years later I was surprised when he told me that was all it was, a prediction. He had faith in me, but had no inside information.

As a Parthian shot, in May 1959, just before leaving Columbia, I sent in my article on renormalization and infinities for publication. This work was mathematically the most difficult thing I have ever done. In it, I relied on a deep result of rigorous mathematics known as the Heine–Borel theorem, which I had learned in a course on pure mathematics at Cornell but never thought I might apply in physics. This paper had some impact, and was expounded a little later in detail in an influential new treatise on quantum field theory. So at least I was leaving Columbia with my flag flying.

7 San Francisco and Berkeley

In June 1959, we flew to San Francisco. That day, we fell in love with the incomparably beautiful city. But Berkeley being our actual destination, in a few days we took a room in a cheap hotel on Telegraph Avenue in Berkeley across the Bay. I checked in at the Radiation Laboratory ["Rad Lab"], and almost immediately was felled by a slipped disc. I could not walk. But this cloud had the proverbial silver lining. An important event would shape my future thinking and my earliest books. This is how I described it in the preface of a recent book:

> Many years ago when I was bed-ridden in Berkeley with a bad back, my wife bought me a present, a paperback copy of Chandrasekhar's 1939 classic, *An Introduction to the Study of Stellar Structure*. She had found it in a bookshop on Telegraph Avenue, and remembered the importance of it from her freshman astronomy course at Cornell. Reading Chandra's book saved me from wasting that awful time in bed, and gave me a permanent sense of excitement that physics and mathematics could deal effectively with something as mysterious as the stars.

This sense of excitement with astrophysics stayed with me, but two years would pass before I took up astrophysics in my own work.

With our Princeton experience behind us, we realized we should live in San Francisco while we were young, without children, and had the chance. On a visit to Berkeley, Murph Goldberger expressed some friendly concern that, by living across the Bay from Berkeley, I would be offending the Berkeley physicists. He need not have worried. My Berkeley colleagues were delighted to have a place to visit in San Francisco.

At that time, Berkeley was the world's leading center of experimental research on elementary particles. The cyclotron had been invented there in 1929 by Ernest Lawrence. With it, charged particles could be accelerated to high enough energy to allow them to break through the electrical repulsion of atomic nuclei, and thereby to probe the structure of the nucleus.

In the following years, accelerators were built at Berkeley with higher and higher energies, and in the 1950s began to produce new kinds of particles that had never been seen before, even in cosmic rays. The climax of accelerator building at Berkeley came in 1954, with the commissioning of the Bevatron, which was designed to accelerate protons (the nuclei of hydrogen atoms) to energies high enough so that antimatter could be created in their collisions. It worked. In 1955, Owen Chamberlain and Emilio Segrè discovered the antiproton in the reactions of protons accelerated at the Bevatron.

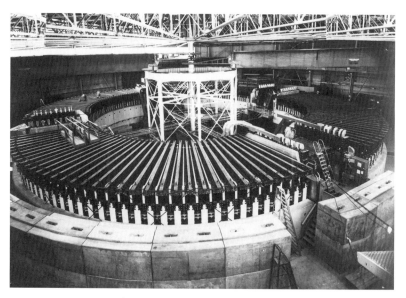

FIGURE 7.1 The Bevatron at Lawrence Berkeley National Laboratory, in 1954

Joining a laboratory with these traditions, I resolved to do something that would be useful to Berkeley experimenters. I soon had an opportunity. Don Glaser had invented an instrument, the bubble chamber, that could be used to study the charged particles produced in elementary particle processes by making their tracks visible. He told me that his bubble chamber group was getting floods of data on the decay into three pions of a particle known as a K meson. I saw how to use the data to test a rule that had been proposed to govern the decay of particles like K mesons, known as the Delta $I = \frac{1}{2}$ rule. A little work showed that Glaser's data confirmed this rule, as I reported in 1960 in my first paper at Berkeley.

Another research group at Berkeley was studying muons (the particles about which Rabi had asked, "Who ordered that?") produced at one of the big Berkeley cyclotrons. Muons decay into electrons and neutrino pairs with a mean life of about 2 microseconds, but if caught in a material of sufficiently high density, such as a liquid or a solid, would sometimes be absorbed by an atomic nucleus before decaying. The absorption of muons in hydrogen was particularly interesting, because this reaction was analogous to the much-studied process of nuclear beta decay, but with the mass of the muon providing much more energy as a probe of the process. Since the nucleus of hydrogen consists of just a single proton, this process could be analyzed without worrying about the effect of the forces between neutrons and protons present in heavier nuclei. However, study of muon absorption in liquid hydrogen was complicated by the fact that a muon caught in orbit around a hydrogen nucleus would often attract another hydrogen nucleus before being absorbed or decaying, producing a proton–proton–muon compound that, with the ready wit for which they are so well known, physicists called a mulecule. All three particles in the mulecule are spinning, and the absorption rate depends in a complicated way on the orientation of these spins. I wrote a paper sorting out all this. When a visiting experimentalist, Valentine Telegdi, gave a talk on muon physics at the Rad Lab, he was kind enough to praise this paper, boosting my spirits and my credit with the Rad Lab.

Telegdi was one of the brilliant physicists who came from Hungary to the United States in the twentieth century. The older generation – Leo Szilard, John von Neumann, and Eugene Wigner – were sometimes called the Martians, on the theory that, if the Earth were secretly visited by Martians, they could be detected by the facts that they were (1) unnaturally intelligent and (2) spoke a language that no one else could understand. In 2005, when I was on a drive through Budapest with Louise, our host pointed out a high school and told us, "This is the school the Martians attended."

I met another physicist in the spring of 1960 who also became a good friend. One day, I was sitting in my tiny office at the Rad Lab, when I received a visitor with a spectacular moustache, who I thought resembled a character from *The Lives of a Bengal Lancer*. It was the Pakistani Abdus Salam, whose work I had studied with care at Princeton. Salam had seen my article proving the validity of the Dyson–Salam method of eliminating all infinities in quantum electrodynamics, and we traded compliments about each other's work.

The most prominent theoretical research group at Berkeley, to which Stanley Mandelstam had gone when he left Columbia, was headed by Geoffrey Chew. The program pursued by Geoff and his collaborators at Berkeley is most often called S-matrix theory. The idea was that we theorists should not concern ourselves with things that cannot be observed, like the quantum fields of particles such as baryons or mesons, but instead should deal only with probability amplitudes, which could be observed, assembled into the S-matrix. Instead of assuming field equations as fundamental principles, the work of particle theory should be based on assumptions about the S-matrix: its unitarity and various symmetry principles, such as Lorentz invariance, charge conservation, and, for the strong interactions also, some approximate symmetry principles, such as isotopic spin conservation. In addition, it was necessary to make an assumption of maximum analyticity. Expressed as a function of Lorentz-invariant energies and momentum transfers, the S-matrix is assumed to be as analytic as possible, with only those poles and branch points required by unitarity. In brief, all features of the strong interactions could be understood as consequences of the general principles of relativity and quantum mechanics.

I thought that the aims of this program were very attractive, if they could be achieved. It is always good to pare down to an essential minimum the assumptions on which our work is based. But I did not think that the S-matrix program could be made to work as a substitute for quantum field theory.

For one thing, in the late 1940s, quantum field theory had scored great successes in understanding electromagnetic interactions. Also, by the end of the 1950s, we had a quantum field theory of weak interactions. This worked well as applied to beta decay in the lowest order of perturbation theory, although it gave unphysical results when extended to higher orders. I did not see how quantum field theory could be part of the fundamental structure of physical theory as applied to the electromagnetic and weak interactions but not the strong interactions.

Also, on a more practical level, I doubted the ability of theorists to deal with functions of several complex variables in implementing the principle of maximal analyticity. I certainly doubted my ability. It seemed to me that, in practice, we would simply have to assume that the analytic structure of S-matrix elements was whatever was provided by quantum field theory.

In the spring of 1960, I had a message from the chairman of the Berkeley physics department, offering me a tenure-track position as an assistant professor. I am sure that this would not have been possible if I did not have the support of Geoff Chew. This offer was what I had hoped for in going to Berkeley, and of course I accepted.

Some time that spring, I received another exciting invitation, to join the JASON group of defense consultants. I had earlier heard of this group at a time when it was first being formed. On a visit to La Jolla to give a research talk at the University of California campus there, I was in the back of a car driven by a UC professor, Keith Brueckner, an accomplished researcher on nuclear and plasma physics. He was trying to enlist his other passenger, Eugene Parker, to join a new group of defense consultants, then known as Project Sunrise. I thought it sounded pretty interesting. It would involve top-secret work on problems important to national defense, and was being convened by leaders in physics, including Murph Goldberger, Charles Townes, and Kenneth Watson as well as Brueckner himself. Parker was the world's leading expert on the physics of the solar corona and wind, so I was not surprised that it was Parker and not me whom Brueckner was trying to recruit.

As it happened, the first summer study of the new group, now called JASON, was to be held at the Berkeley Rad Lab in the summer of 1960. Perhaps because I was already there and already had an Atomic Energy Commission "Q" clearance (which was routinely required for scientists at the Rad Lab), I was asked to join. The nation was still reacting to the dramatic flight of the Soviet spacecraft Sputnik in 1957. This had generated worries about American technological capacity, and motivated an enhanced willingness to support scientific research. In the spirit of the times, the Department of Defense had given a broad charter to JASON to explore new technologies of possible military significance. It had been founded in January 1960 with administration provided by the Institute for Defense Analysis and funding by the Advanced Research Projects Agency. Summer of 1960 was the beginning of my long association with the JASON group, a major part of my working life for over a decade.

Later in the summer of 1960, I returned to Rochester for a few weeks. A few years earlier, a series of international conferences on high-energy nuclear physics had been initiated at Rochester by Robert Marshak. I was too junior to be invited as a regular participant, but I had an invitation to attend the conference as one of the scientific secretaries, who would convert transcripts of the talks given at the meeting into publishable texts. It was a terrible job. I learned that what people actually say in public talks is very different from coherent publishable language. But this gave me a chance to listen in to talks on the latest results in particle physics theory and experiment.

I began my university teaching in the 1960–61 academic year. I was assigned to teach a course on vector algebra taken mostly by engineers. Vectors are quantities like wind velocity or magnetic fields that have

FIGURE 7.2 The author as a young professor

both magnitude and direction. You can add vectors to make other vectors, like adding the magnetic fields produced by two different magnets, or you can multiply vectors by other vectors. I understood vector algebra as an application of an invariance principle, the principle that the laws of nature seem the same to two observers if one observer's frame of reference is rotated compared with the other's. Vectors are defined by the way they change when we rotate our frame of reference. The engineering students in my class explained to me that they were disappointed with this approach. They wanted to learn vector algebra as a tool for solving engineering problems, and were no more interested in its rationale than they would be interested in the process used to produce the paper on which their textbooks were printed. From this experience, I realized the importance of teaching the particular use of a mathematical method. I would have had to learn some engineering to do this, but in physics, teaching the insight has been valuable. I also realized I loved teaching.

In 1960–61, I wrote some papers on weak interactions, some with Gary Feinberg and one with Jeremy Bernstein. It was interesting, but

I did not want my work to be confined to the weak interactions, and was feeling restless. I picked up some books on a branch of mathematics, complex analysis, and saw how a rigorous theorem set out by the mathematician Herglotz could be applied to improve a 1958 result of the Russian physicist I. Ya. Pomeranchuk. Pomeranchuk had shown that, under some reasonable assumptions, the total collision rate at which particles would interact with any target would be the same as for the corresponding antiparticles. I used the Herglotz theorem to show how the same result could be derived under more general assumptions.

That spring, I had an invitation to come down to Cal Tech to talk about this paper. It was a traumatic experience. In the audience were Murray Gell-Mann and Richard Feynman. Murray asked detailed questions, probing for any weakness in my arguments. Feynman was even more difficult. For some reason, he found my reference to the Herglotz theorem amusing, as if I had made it up to sound impressive. Like everyone else, I admired Feynman's brilliance and originality, but did not find much to like about him personally until much later. Indeed, whenever I went down to Cal Tech to give a talk, Louise would say I seemed nervous. It was dread of the going-over that Murray and Dick had in store for me.

FIGURE 7.3 Murray Gell-Mann and Richard Feynman in 1957

JASON went on a field trip that spring. Members were to take commercial flights to Washington, assemble at Andrews Air Force Base (now known as Joint Base Andrews), and take a Military Air Transport Service flight down to the US Naval Base at Key West. On this flight, JASON members kept pestering our administrator, David Katcher, asking what would be our simulated officers' rank at the naval base. Exasperated, he told us that we would be simulated non-commissioned petty officers.

The first day in Key West, we participated in an exercise in anti-submarine warfare, a recurrent preoccupation of the JASON group. We variously signed up to board aircraft, surface ships, and a diesel submarine. Wanting to be in the open air, I opted to spend the day on a destroyer escort. Freeman Dyson was a shipmate. All vessels and aircraft spent the day searching for the submarine, without success. We came away with some understanding of the importance of submarines to the national defense – a submarine is very difficult to find.

That evening a group including Keith Brueckner, Don Glaser, Francis Low, and myself went to check out the Mardi Gras. We arrived just as a pretty girl started dancing on the bar. Don Glaser had the idea of photographing Brueckner, Low, and me through the dancer's legs. Unfortunately, a bouncer took Glaser's camera from him and smashed it with his bare hands. It was a great pity because this camera contained all the photos that Glaser had taken during his trip to Stockholm to receive the Nobel Prize the previous December for his invention of the bubble chamber.

It was in the spring of 1961 that I began my active involvement with cosmology. I felt that particle physicists could make useful contributions to cosmology. Chandrasekhar had explained why the stars shine. But there was the whole cosmos to think about. Reading a book on cosmology by the Austrian–British astrophysicist Herman Bondi suggested a problem that might provide me with an entrée to the field.

Bondi described a classic difficulty of cosmology, known as Olbers' paradox. If we supposed that the universe were infinite, static, and everywhere pretty much the same, then no matter how far apart the stars might be, every line of sight when extended outward would ultimately reach the surface of some star. So why is the sky not covered with stellar light? The answer, as Bondi explained, is that the universe is not static but expanding. This has two effects. First, because distant galaxies and the stars they contain are moving away from us with a speed that increases with distance, the light we receive from distant and more distant sources is increasingly shifted toward the red low-frequency end of the spectrum, and hence toward lower energy and less detectability. (This of course is how we know that the universe is expanding.) Second, the observed rate of expansion suggests that the expansion began about 10–20 billion years ago, so that we only see things no farther than 10–20 billion light years away, in which case there would not be time for starlight to have reached us along most lines of sight.

It occurred to me that, although the expansion of the universe resolved Olbers' paradox for light, it might not solve it for neutrinos. Neutrinos were supposed to be like electrons, in that they carry a conserved quantity known as lepton number. The redshift due to cosmic expansion could wipe out the energy of neutrinos from distant sources, but not their lepton number. Also, it was widely speculated at the time that the universe is infinitely old, with the current expansion only the latest phase in an infinite cycle of expansions and contractions. I reasoned that in that case the neutrinos emitted by stars in past cycles could still be with us. Neutrinos are like electrons in another sense. They satisfy Pauli's exclusion principle, according to which no two neutrinos can be in precisely the same state. Therefore, neutrinos left over from early cosmic cycles could not all sink to the state of zero energy, but would have to form a distribution with energies ranging from zero to some maximum, just like the "Fermi sea" of electrons in any metal.

In addition, I even thought of a way of detecting these neutrinos. The isotope hydrogen-3 undergoes a radioactive beta decay into helium-3, plus a neutrino and an electron, with a total released energy that is unusually small for a nuclear beta decay. Because of the exclusion principle, a Fermi sea of cosmic neutrinos would block the emission of neutrinos of the lowest energy, so that electron energies could not quite reach their expected maximum, while a Fermi sea of cosmic antineutrinos would lead to a few events in which antineutrinos were absorbed from the Fermi sea in place of neutrinos being emitted in the beta decay, giving the emitted electron an energy slightly larger than the expected maximum.

We know now that this whole picture is wrong. Whether the universe is infinitely old or has a finite age, we have other information that changes the preexisting understandings. Ever since the 1965 discovery of a cosmic background of microwave radiation, we have understood that, near the beginning of its present expansion, the universe went through a phase when neutrinos and radiation and electrons and almost everything else shared a common rapidly decreasing temperature. This has definite consequences for the cosmic distribution of neutrinos. As a result of the expansion of the universe, neutrinos now have a temperature of about 2 Kelvin (degrees Celsius above absolute zero), too low to be detected in the decay of hydrogen-3. I did not know about this in 1961, and published a paper on what I called "The Neutrino Problem in Cosmology."

Conceiving of physics as a worldwide enterprise, I thought of taking a year abroad to become acquainted with physicists elsewhere. I applied for and received an Alfred Sloan Fellowship, so I could pay my own salary and could pretty well take my pick of physics departments to visit. England was a natural choice; it would avoid language difficulties. I wrote to Abdus Salam, who was then the head of the theory group at the Imperial College of the University of London, proposing a visit in the 1961–62 academic year. Salam agreed, so that was settled.

Louise then had a clever idea. The Earth is round, so we could get to London from San Francisco by going west instead of east. We could see something of Asia and then travel through the Soviet Union on the way to England. When our time abroad was coming to an end, we could continue westward again, over the Atlantic and the US back to California.

The best of the idea she had not known. In those days, most world airlines coordinated their international air fares through the International Air Travel Association, so that airlines would accept tickets even if they had been issued by a different airline, and several airlines offered flexible round-the-world opportunities. For $1,268 each, we bought two tickets that would take us all the way around the world on any routes and airlines we chose, as long as we kept going more or less in a westerly direction. For me, the advantage would be the chance to introduce myself to colleagues the world over. I saw the usefulness of such a step, having seen many foreign visitors to physics departments in the US.

But first there was the summer of 1961. In the early summer, we spent a few weeks in Madison, Wisconsin, where there was a summer program in theoretical physics at the University of Wisconsin. I spent much of that time talking with Jeffrey Goldstone, a visitor from England, about a new idea in theoretical physics, the idea of broken symmetry. I had been interested in this idea since I had heard about it at a lunch that spring, from a visitor to Berkeley who later became our dear friend, Sergio Fubini. This idea was going to be at the center of my own work and the work of many other theorists over the next few decades. It can't adequately be explained without equations, but it is important enough for me to try to give some sense of the excitement and difficulty of this work.

Much of the progress that had been made in recent years had to do with principles of symmetry, principles that grouped particles into families with equal or related properties. There was isotopic spin symmetry, which grouped the proton and neutron into a family and grouped various other nuclei into other families. There was the more extensive symmetry introduced by Gell-Mann and Ne'eman known as the "eightfold way," which grouped protons and neutrons together with other particles that, in accelerators, are produced in octets.

Nature obeyed these symmetries in some cases, but not in others, and nobody knew why. The new idea that Fubini conveyed to me was that there may be many other symmetry principles obeyed by the equations of physics, which are broken in the sense that they are not obeyed by the solutions of the equations that govern observed properties of the elementary particles. Not being obvious in our observations, there might be many broken symmetries waiting to be discovered. I thought that this was interesting.

This speculation about broken symmetry began with a study of the currents that play a role in the theory of weak interactions – something like the role of electric currents in electrodynamics. There is a deep relation between symmetry principles and the existence of currents that are conserved. This was first noted by Emmy Noether in 1918. A current is said to be conserved if the amount of something in any volume increases or decreases at the same rate as that quantity is carried into or out of that volume by the current. For instance, the electric current is conserved. The amount of electric charge in any volume increases or decreases at the same rate as the charge is carried into or out of that volume by the electric current. And it has been understood since the work of Marshak and Sudarshan, and of Feynman and Gell-Mann, that the weak interactions are produced by two kinds of currents. One is a "vector current," which has properties very similar to the electric current. In particular, Feynman and Gell-Mann had argued that this vector current is conserved, like the electric current. The other is called an "axial vector current," which behaves differently in several respects. In particular, it did not seem to be conserved.

In 1960, Yochiro Nambu noted that, because of the presence of the "induced pseudoscalar" term in the beta decay interaction (the term that I had described in my PhD thesis), there was a possible cancelation that would allow the axial vector current to be conserved, provided that the pion had zero mass. The pion is not massless, but it is by far the lightest of all the particles that participate in strong nuclear forces, with about one-seventh the mass of the proton or neutron, so Nambu was really suggesting that the axial vector current is approximately conserved.

For this approximate cancelation, it would only be necessary to satisfy a certain relation between the strength of the axial vector current contribution to neutron beta decay and the strength of the interaction in which pions are created or destroyed by neutrons and protons. This relation had already been proposed in 1958 by Goldberger and Treiman on the basis of very shaky arguments. The Goldberger–Treiman relation was known to agree with observation, and now Nambu had given a plausible explanation for it.

The symmetry underlying conservation of the axial vector current of weak interactions would be a kind of isotopic spin symmetry, but "chiral," in the sense that the proton and neutron states spinning to the left and right are grouped into two different families. No such symmetry is actually observed in the interactions of neutrons and protons, so this symmetry would have to be broken – present in the equations of the theory, but not in the phenomena.

At Madison, Goldstone explained to me that, although he had no problem with Nambu's broken approximate symmetry, he doubted that there was room in physics for any broken exact symmetries. His calculations in several field theories showed that broken exact symmetry would imply the existence of particles that are not light, like the pion, but massless and spinless. They came to be called "Goldstone bosons." No such particles were known.

When a theory leads to the prediction of a new type of particle that has not yet been observed, if the mass is not predicted, theorists can always cross their fingers and guess that the new particle is just too heavy to have been produced at existing accelerators. But that option is not open if the particle predicted is massless. So it seemed that, if Goldstone were right, then there could not be any broken exact symmetries. It took a few years before we learned that this is almost but importantly not quite correct.

We left Wisconsin or the 1961 JASON summer study, which was to be on the Bowdoin campus in the cool and pretty town of Brunswick, Maine.

The work I did that summer for JASON forced me to learn about magnetohydrodynamics, the theory of various kinds of wave in ionized gases permeated by magnetic fields. I made a modest advance in this theory, publishing a paper on it the next year. Through defense work at JASON, I had the experience, unusual for a theoretical physicist, to learn about phenomena of some use in the real world: plasma physics, hydrodynamics, sound propagation in the ocean, radar propagation, and much else. Yet those bodies of practical knowledge turned out to be invaluable to me as a theorist. I could not have written my later books and papers on astrophysics and cosmology without this background.

In much of this, my mentor throughout the Berkeley years was Kenneth Watson, a senior professor. Ken took Louise and me on a visit to Los Alamos, and arranged for a consultantship for me at Convair in San Diego. We became close friends with Ken and his wife Elaine. It was a good summer in other ways. John Wheeler was an esteemed and deeply knowledgeable presence.

The summer over, we returned to San Francisco, and packed up. We booked on Pan Am flights from San Francisco to Honolulu and then on to Tokyo and Hong Kong, a route designated as "Pan Am Flight Number 1." Always heading west on jet planes, we set out to circumnavigate planet Earth.

8   East to London

I had written in advance to the physics department at the University of Tokyo, letting them know I would like to visit. After a couple of sybaritic nights in Honolulu, we flew to Japan. Physicists in Tokyo invited me to give a seminar on cosmic neutrinos. After my talk, I was taken to pay my respects to the august Sin-Itiro Tomonaga, one of the founders of modern quantum electrodynamics. Tomonaga was among those great men in whose lives high position in administration eventually takes the place of research. I respected his position, of course, which certainly honored him and his work. But greeting Tomonaga caused me to reflect. I realized that, whether or not I could make contributions of importance, the path of administration was a road I myself would leave untraveled.

A member of the physics department, Nobiuki Fukuda, kindly undertook to take us around Tokyo. He offered us all of his time and his truly genial companionship.

Another member of the Tokyo physics department, Hiroomi Umezawa, was known to me through his book on quantum field theory, which I had read at Columbia. Umezawa's wife was a student of the traditional Japanese tea ceremony. It was gracious of the Umezawas to invite us home to witness a performance of this ancient custom. At their apartment, Mrs. Umezawa performed the ceremony for us artistically. Since the weather was hot, tea was served cool. It was a thick green liquid. At several moments during the ceremony, we were prompted to praise the tea formally, a part of the ritual. One says, "Very good. Very good." Louise asked, "What if we really like the tea?" After a pause, Umezawa replied, gravely, "Sometimes misunderstandings arise."

After our week in Tokyo, we took an astonishingly fast and comfortable bullet train to Kyoto, the ancient capital. At the university in Kyoto, I spoke again about cosmic neutrinos. I was introduced to Hideki Yukawa, who in the 1930s had famously predicted the existence of a particle he called the "meson," because of mediating interactions between protons and neutrons within the nucleus of atoms. This was thirteen years after the discovery of the proton and three years after the discovery of the neutron. In 1949, Yukawa was awarded the Nobel Prize for predicting the pi-meson, or so-called pion.

After Japan, Pan Am Flight No. 1 took us to Hong Kong. As we settled in for the flight, a woman passenger who recognized us asked, "What do you kids do, just clip bond coupons?" In fact, we had become what we jokingly referred to, between ourselves, as "physics bums." That is, we were living to some extent on the modest honoraria I received for physics talks I gave en route. I would give a talk, my hosts would hand me a couple of hundred dollars in foreign money, and we would get back on Pan Am No. 1.

I had written ahead to physicists in Hong Kong, and, as in Japan, I was invited to speak. The physics department of the University of Hong Kong was housed in a grand white stucco British colonial building on a slope of Victoria Peak. It had high ceilings and huge windows. I made friends with a dignified and handsome young theorist, Chan Hong-Mo, who later would play a significant role in the early days of string theory. Hong-Mo always wore a dark-gray bespoke suit, spoke the purest Oxford English, and was very debonair. He recently reminded me that the three of us had chatted happily about poetry during our stay in Hong Kong, and sent me a copy of his own translations of classic Chinese poetry. I have read some Chinese poetry, translated by Ezra Pound and other well-known poets, but I have to say I found Hong-Mo's translations best of all, more moving and more lyric.

We traveled on to Singapore, and this would turn into a fateful visit for me – I mean fateful for my future as a physicist. We boarded a bus to take us from the airport into town. We watched the driver load all our luggage into a hold in the rear of the bus. But when the bus stopped at our hotel, the Raffles, and our luggage was placed on the ground, a small suitcase of mine was missing. I was distraught. The suitcase contained a loose-leaf notebook in which, over the years after Cornell, I had put notes on virtually everything I was learning in physics. That is why I had insured that suitcase for $700, and why the insurer had agreed to insure it.

The authorities were very kind and took me on a tour of low dives. We were in the bus driver's home as well. But I never got that suitcase back. The English branch of the insurance company paid the $700 in the end, but I was glad of the loss. Losing the notebook turned out to be a boon. Over time, I would have to rethink much of what I had put into it, with a far more sophisticated perspective.

And there was a further benefit. I think it was then that I began to try to speak without notes. Increasingly, I tried to lecture on my feet, forgiving myself the not infrequent mistake on the blackboard. In debates and at conferences, I tried to formulate my responses to what my colleagues were saying as they were saying it.

In Bombay, I gave a talk at the Tata Institute, and made friends with Yash Pal, an expert on cosmic rays. The importance to fundamental physics of cosmic rays in the days before the great accelerators is an untold story, but it cannot be exaggerated. It was in the happenstance of cosmic rays that we first saw certain of the elementary particles.

Pal very hospitably invited Louise and me home to dinner. I had always liked Indian food, but this was the first time I had a chance to try Indian home cooking. Mrs. Pal served us a delicious eight-course dinner.

While in Bombay (now called Mumbai), I was taken out to the Bhabha Research Center in the suburb of Trombay, to visit the CIRUS research reactor supplied to India by Canada in 1954. Looking down into the reactor's pool of heavy water surrounding its uranium core, for the only time in my life, I saw the blue glow of Cerenkov radiation. This radiation is something like the sonic boom produced by a supersonic airplane, but it is an optical boom, light produced by a charged particle traveling faster than the speed of light in the medium through which the particle moves. Of course, nothing can travel faster than the speed of light in empty space, but the speed of light in water (ordinary or heavy) is only three-quarters the speed of light in vacuum, and many of the electrons emitted in the radioactive decay of fission products in a nuclear reactor are traveling faster than that.

Our last stop in India was New Delhi. While waiting for our visas to Russia, I began to show the classic symptoms of what travelers know as "Delhi belly." I had to cancel my talk at the university and, regrettably, give up our side trip to Agra to look at Shah Jahan's Taj Mahal.

I gathered that the Russians did not want an American physicist to come chatting with their physicists. So I went to American Express to make new plans for the rest of our trip. Instead of a flight to Tashkent and then on to Moscow, we would get to London via Tel Aviv, Istanbul, and Athens. There had been no chance to get in touch with physicists along this route, so this part of our trip would be strictly for sightseeing, although I thought I might get in touch with Israeli physicists when there.

In Israel at last, it was a relief to be back in a country that was western, cool, and clean, without beggars. Fragrant orange trees lined the path from the airport to the taxis. Coming up to our room at the Dan Hotel, we saw from our window the brilliant blue of the Mediterranean.

Then, coming down to the lobby, we unexpectedly encountered the physicist Larry Wilets, a good friend who had been hospitable to us, with his wife Dulcie, in Copenhagen. He was visiting the Weizmann Institute of Science in Rehovot, and arranged for me to come out there to meet Israeli physicists and give a talk.

While in Israel, I began to think about founding and directing a winter school in Jerusalem. The plan was to invite distinguished lecturers each year to expound the formalisms in a current problem in theoretical physics. I hoped students and scientists from all over the world would attend. A special hope was that Islamic physicists would participate.

Of course Jerusalem is of signal importance to many faiths that acknowledge their Judaic roots. Jerusalem is, and has always been, even when inchoately, the capital of the Jewish nation. For thousands of years, at every Passover dinner, every Jew, from father to son, had said, "Next year in Jerusalem! Speedily, speedily, in our day, soon." I am an atheist and do not believe in miracles, but suddenly in 1948 it was "this year in Jerusalem." And it had happened "speedily, speedily, in our day." I refrain from calling it a miracle, especially since the lives of brave young soldiers were given to secure it, deep in the West Bank as it is, but it was surely miraculous.

In a few years, I would return to found the Jerusalem Winter School in Theoretical Physics.

With tourist stopovers in Istanbul and Athens, we arrived finally in London. I found a fine theoretical physics group at Imperial College. The head of the group, and its most distinguished theorist, was Abdus Salam, whom I had met briefly in Berkeley, and whose work I had carefully studied at Princeton. His coworker and contemporary, Paul Matthews, had made significant contributions to quantum field theory. Ray Streater was an expert on the mathematically rigorous aspects of quantum field theory, and a collaborator with Arthur Wightman, whose course on advanced quantum mechanics I had taken at Princeton. In some ways, the most interesting person was Tom Kibble. He was tall, quiet, and blinked a good deal, almost as if cast as an English professor in a play. I had long been impressed with his work applying the mathematics of topology to cosmological fluctuations. Tom was knighted in 2013 for work on broken symmetry, which would later become a subject of particular interest to me.

At this time, I began work on a research project that would occupy me for the next several years, but which turned out to be a failure. I am going to describe this, if only briefly, because it may provide a useful contrast to so much that is written about the history of science, in which the reader is shown only one success after another.

FIGURE 8.1 Abdus Salam, in 1955

One of the greatest blocks to progress in fundamental physics in the 1950s and 1960s was the strength of the strong nuclear forces. We knew in principle how to calculate the probabilities of reactions by adding contributions from various scenarios in which particles are emitted and absorbed by the particles in the initial and final states and by each other. In relativistic quantum field theories, there is always an infinite number of these scenarios, each of which could be represented by a Feynman diagram, so exact calculations are impossible. However, the interactions in quantum electrodynamics in which particles are created or destroyed are rather weak, so that for each additional complication, the contribution of one scenario to the probability is suppressed by a small factor, of order $1/137$, and good approximate results can be calculated by taking account only of a few simple scenarios. This is known as perturbation theory. In dealing with strong nuclear forces, the corresponding factor is not small, like $1/137$, but roughly one, so perturbation theory gets you nowhere.

I focused on one annoying aspect of strong forces, that they can produce bound states. For example, the nuclear force between a proton and a neutron is strong enough to bind these two particles in a composite, the nucleus of heavy hydrogen known as a deuteron.

This is a clear sign of the failure of perturbation theory in proton–neutron interactions.

Suppose a scenario in which a proton and neutron exchange a meson (as Yukawa had theorized), two mesons, or any number of mesons, but all you arrive at, in your calculation, is just a proton and a neutron and a lot of mesons, no deuteron.

I could see a way, in the nonrelativistic case, where the neutron and proton directly exert forces on each other like the gravitational force between planets, in which it would be possible to rewrite the theory so that the deuteron appeared as a "quasi-particle," a fictitious elementary particle, to be sure, but one that would not change the physical content of the theory. The neutron–proton force would have to be weaker in the rewritten theory, to avoid the emergence of a second deuteron as a bound state. The upshot might be to make it possible to use perturbation theory to calculate neutron–proton scattering.

This actually works for the nonrelativistic case, but that is a case in which the method is not needed, because accurate calculations can be done in the nonrelativistic case using modern computers without the use of perturbation theory. I never saw how to adapt this method to relativistic quantum field theory, where it might make a difference, and no one else became interested in it. This was not an example of the physics establishment being too unintelligent to adopt a new idea. Rather, the world of physics was intelligent enough to see that this idea was going nowhere.

After a month or so in London, I received a message from the consular section of the Soviet embassy. Our visas had arrived, and we were welcome to arrange with Intourist for a visit to the Soviet Union. I believe these obviously useless visas were just a bit of diplomacy pasted over irrational guardedness. I politely replied that the invitations, unfortunately, had come too late.

I gave invited talks, Louise accompanying me everywhere, at Birmingham, Cambridge, Edinburgh, Glasgow, Manchester, and Oxford.

In Cambridge, I visited the Cavendish Laboratory, where in 1897 J. J. Thomson had discovered the electron, and in 1932 James Chadwick had discovered the neutron. At the Cavendish, I met Fred Hoyle and Willy Fowler, who with Margaret and Geoffrey Burbidge had figured out how elements heavier than helium are synthesized in stars.

Fowler showed me the bas-relief of a crocodile on the wall of the laboratory, supposedly a symbol of Rutherford, who in 1919 had come to direct the Cavendish after discovering the nucleus of the atom at Manchester. The crocodile was part of what had inspired one of my books, *The Discovery of Subatomic Particles*.

In London, my interest returned to broken symmetry. I saw how to give what I thought was a more convincing proof of Jeffrey Goldstone's conclusion that the spontaneous breaking of exact symmetries always leads to the existence of massless particles. Salam became interested, and we had many discussions about it. When I took Louise touring in Edinburgh, I had a chance to discuss broken symmetry with Peter Higgs, who later did very important work in this area, along the same lines as Kibble's. These early conversations turned out to be useful.

At the Edinburgh faculty club, Higgs introduced me to single malt Scotch whisky, for which I am forever grateful. I think it was Higgs who liked to show up in a kilt. Louise kidded him about "that skirt."

I planned that, in the spring, we would leave London and fly to Naples, to work our way north through Rome, Florence, and Venice, winding up in Geneva, where in early July, the Conseil Européen pour la Recherche Nucléaire (CERN) would be hosting the 1962 "Rochester Conference" on high-energy nuclear physics, to which (at last) I had been invited. It would be a great event.

Before leaving London, I hurriedly finished the paper on broken symmetry, with Salam as coauthor. It occurred to me that I had picked up the idea that broken symmetry entails massless particles from my time in Madison with Goldstone, and one of the proofs of this result in our paper was largely based on Goldstone's work. I cabled Goldstone describing the paper, and inquiring whether he wanted to be named as coauthor. (Yes, in those days, there was no email, and people sent each other telegrams.) A few days before our departure from London, I received a telegraphic reply from Goldstone: "Make me author." So we did.

In Naples, I visited the research group of Eduardo Caianiello, whom I had met at Columbia. Their building was a short walk from the zoo at the Mostra D'Oltremare, the overseas exposition erected in the days of Italy's African empire. On one occasion, outdoors at the zoo café, I had lunch with another visitor, Norbert Wiener, a great mathematician and the founder of cybernetics. A gaggle of little school girls stopped at our table, fascinated by Wiener's wild beard. Wiener spoke to them impressively in what I took to be the Neapolitan dialect.

After Naples, we went on to Rome, and found it full of friends. Francis and Natalie Low were there, and had us over to their Rome apartment. Francis was now a professor at MIT and a JASON member. He was one of the wisest theorists anywhere. While at the University of Illinois in 1956, Francis had developed an approximate theory of meson scattering by nucleons, collaborating with Geoffrey Chew, who in 1962 was my colleague and the leading theorist at Berkeley, Earlier, in 1954, Francis had written a paper with Gell-Mann, describing how the effective value of electric charge varies with the energy at which it is measured. When this calculation was extended later, to the theory of strong nuclear forces, it became crucial in the development of the Standard Model of elementary particles. In a 1983 festschrift for Francis, I said that the Gell-Mann–Low paper was one of the very few whose journal reference I knew by heart, and the few others whose references I knew by heart were by me. The Lows and Weinbergs were to become very close friends when we lived in Cambridge.

I made a new friend in Rome, Nicola Cabibbo, a theorist just a little younger than my generation. He had grown up in Rome, and took me on a tour of its surroundings: the ancient grounds of Tivoli, after which the park in Copenhagen was named, and Frascati, site of a physics laboratory and a cherished vineyard. Cabibbo became famous later when he worked out a theory uniformly encompassing both ordinary weak interaction processes, like nuclear beta decay, and the weak decays of more recently discovered "strange" particles.

Cabibbo's theory involved one new numerical quantity, an angle that everyone called the "Cabibbo angle." Everyone but Murray Gell-Mann. Murray evidently thought that Cabibbo's theory was implicit in his own work, and so he always referred to this quantity as "that funny angle." Cabibbo told me that he was thinking of changing his name from "Cabibbo" to "Funny."

I don't recall meeting any physicists in Florence, but we did take a side trip to Pisa, where I gave a talk at the Escuela Normale Superiore at the invitation of the theorist Luigi Radicati. Radicati was soon to be a leader in the effort to combine multiplets of particles of different spin, a program that had some (limited) success when applied to phenomena in which relativistic effects are not large. While in Pisa, I paid my respects to the Leaning Tower, Galileo's accelerator. Radicati took us out to a restaurant that specialized in wild game.

Marvelous Venice was our last stop before Geneva. At last, Geneva. We would be there for a few weeks, waiting for the beginning of the 1962 "Rochester" Conference on high-energy nuclear physics. Several days a week, I would take a bus out to CERN, in Meyrin, a suburb of Geneva. At that time, experimental research at CERN centered on use of a new accelerator, the Proton Synchrotron. It was huge in comparison with the tabletop apparatus used by Rutherford to discover the nucleus of the atom at Manchester, but tiny by the standards of CERN's later accelerators, a mere 200 meters in diameter. Today's largest accelerator, the Large Hadron Collider at CERN, is an underground ring with a diameter of 8.6 kilometers – a little over five miles. In 1962, the Proton Synchrotron put CERN in the big leagues of experimental research on elementary particles, along with the Rad Lab in Berkeley and Brookhaven on Long Island.

In late June, Geneva began to fill up with physicists in town for the high-energy physics conference. Many were friends. With Gary Feinberg and Jeremy Bernstein, we drove out to Père Bis, a famous restaurant across the border in France on the shore of Lake Annecy. Gary was then into wine lore, and insisted on ordering a bottle of Richebourg burgundy, which cost more than the rest of the meal.

Another friend, Bruno Zumino, arrived from New York. I had met Bruno when at the Columbia physics department I heard him explaining a theorem, the CTP theorem, to T. D. Lee. This theorem states that any quantum field theory, however constructed, will always respect a symmetry under the combined interchange of particles with their corresponding, oppositely charged, particles (C), together with reversal of the direction of time (T), and the parity transformation that reverses all spatial directions (P). The experiments suggested by Lee and Yang, which showed that P symmetry was not respected by the weak interactions, had also incidentally showed that invariance under TP transformations (which reverse the sign of all space and time coordinates) is not respected either. And so, according to the CTP theorem, the weak interactions could not respect invariance under C. If they did, then CTP invariance would be violated along with TP invariance.

All this was then new to me.

Bruno and his then wife, Shirley, took us out to dinner at a wonderful restaurant in Geneva, and we four became fast friends.

At last the Rochester Conference of 1962 got under way. I gave a talk in a session of "contributed talks" that were outside the regular sessions of invited talks. I chose to describe my ideas about using the introduction of quasiparticles to allow calculations in theories of strong interactions. My talk met with a well-deserved lack of excitement. It was a bad choice of topic. I should have spoken about the paper on broken symmetry with Goldstone and Salam, which turned out to be important.

But it was a happy time. I recall that, at a café outside our hotel, Oppenheimer, who must scarcely have remembered me from Princeton days, nevertheless graciously stopped at our table to greet me and chat for a moment. He had read and understood some of my work, as he read and understood everything. When he walked on, a foreign diplomat in exotic costume came up to our table to ask, excitedly, "Was that Oppenheimer?" I was almost as excited by the event as he was.

During another coffee break at the meeting, I came across Geoff Chew, who told me that the physics department at Berkeley had voted to promote me to associate professor. This was a meaningful step-up. Associate professors have tenure. So I thought that at last Louise and I could see ourselves permanently settled in Berkeley. (As it turned out, our future was to be very different.)

After the conference, we were to fly through Lisbon to New York to see our parents. The stopover in Lisbon was a long one. At a round table outdoors at the Lisbon airport, Oppenheimer was presiding over a discussion among physicists waiting for connecting flights. In talking over the conference, they were expressing some regret. Most attention at the CERN meeting had been directed to strong nuclear forces, but no one was offering a serious theory of these interactions, in the sense that quantum electrodynamics is a theory of electromagnetic interactions. Instead, the emphasis was on saying things about reactions at very high energy that would be valid in any theory. There had been a good deal of attention to the "S-matrix program" pursued by Geoff Chew, Stanley Mandelstam, and their collaborators at Berkeley, which I have already described. It was an attractive idea, but had scored no significant successes. Very little attention at the conference had been given to the weak interactions. There was a good working theory of nuclear beta decay, but it made no sense when pushed beyond the simplest approximations, and no one was offering any ideas about how to fix it. We did not know it then, but the sense of frustration that we felt sitting around a table outside the Lisbon airport would be replaced in a little more than a decade by a well-earned feeling of triumph.

We stayed in New York for just a few days, touching bases with our families, and then flew on to San Francisco, still on Pan Am No. 1. We had gone around the world in a bit under 280 days.

9 Berkeley

By the time we returned to California, Louise and I were tired of travel, so it was fortunate for us that the 1962 JASON summer study was to be held again at the Rad Lab in Berkeley. The JASON staff rented a house for us in the hills above the University of California campus, which we planned to use in off-hours as a base for our search for a more permanent rental.

My Princeton PhD advisor, Sam Treiman, had joined JASON, and that summer we collaborated on a paper, "HBT on a Wake." The title is unclassified, and though I am not sure that the paper itself has ever been declassified, it is no secret that HBT is an acronym, well known to physicists, standing for Hanbury-Brown and Twiss. They were the inventors of a new kind of interferometry. In general, interferometry is the measurement of the properties of a distant source by combining the signals that arrive from it at a detector by different routes for instance, capturing the light of a star in two different telescopes and sending both light rays to the same film or eye. Usually these observations are sensitive to the difference in the phase of the two signals arriving on different paths; this is the technique, for instance, that was used in the first measurement of a star's diameter. Hanbury-Brown and Twiss had shown that interferometry can also work for a signal that is fluctuating randomly when it is amplitudes rather than phases that are being compared. I don't think that our paper had any value for national defense, but once again JASON was providing me with a valuable education.

We found a woodsy house in the Berkeley hills at 1390 Queens Road. One had to rent. Prices in Berkeley even then were beyond the purse of a mere professor. Our friends, Art and Roz Rosenfeld, had managed to find an affordable big place, a house in the grand style.

A maid in a frilly white apron would open their door. Roz told us that they had got the house cheap because it was on the San Andreas Fault, and likely to be shattered in an earthquake. On reflection, I think that was just to play down some actual wealth they enjoyed.

During that summer, T. D. Lee briefly visited Berkeley, so I took the opportunity of inviting him home to dinner along with Sam and Joanna Treiman. It was good to see them all again. I felt a little more grown up than when I had previously seen T. D. or Sam. It is possible I was the reason for T. D.'s visit. During that evening, T. D. took me aside and asked me if I would be interested in returning to Columbia. I politely and noncommittally told him that we wanted to stay in one place for a while. In fact, at that time, looking for a big enough house for ourselves, we were not interested in trying to squeeze ourselves into a Manhattan apartment. And after life in San Francisco and London, we knew that there was life beyond New York.

In 1962, I published an article with Salam and Goldstone, which turned out to be important, though needing some correction. We had concluded that a massless spin-zero particle would appear for each broken symmetry. But in 1964, several groups working independently (Robert Brout and François Englert; Peter Higgs; and, separately, Gerald Guralnik, Carl Hagen, and Tom Kibble) showed that, in Yang–Mills theories, our thinking could produce a finite mass.

During the years at Berkeley, I began to reconsider how I understood some of the theories of physics – not whether they are true, or why we think they are true, but why they actually are true. I was powerfully motivated in this by the task of teaching. In teaching a theory like Maxwell's theory of electromagnetism, Einstein's theory of gravitation, or Dirac's theory of the electron, the professor can simply lay out the equations of the theory for the class. But this leaves us (or should leave us) with a hollow feeling.

Electromagnetism and gravitation and electrons are all very interesting, but what is really important is to know how these theories will ultimately fit into our understanding of the laws of nature that underlie them, and for that we cannot simply say that these are the theories that have been borne out by experiment. And we cannot rely on the views of the great physicists who gave us these theories. The rationale for physical theories changes with time, even if the theories do not. We know they work, and then we find out why. Indeed, we should not be content with understanding Maxwell's, Einstein's, or Dirac's theories simply in the way they did. Much progress has been made since their theories were first proposed.

I have said that I was dissatisfied with the way that general relativity had been taught at Princeton. In that course, and in standard textbooks, general relativity was understood geometrically, as a theory of curved space-time. In teaching the theory at Berkeley, I gave up geometry as a starting point, and based the subject instead on a physical principle that Einstein had called the principle of equivalence of gravitation and inertia: The inertial force that we feel when we are suddenly accelerated is indistinguishable in its effects from the force of gravity. Since inertial forces act just like gravity, we can arrange that they cancel. That is, in any gravitational field, it is always possible to find a suitably accelerated frame of reference in which inertial forces cancel gravitational forces, and at least in a small region there are no effects of either gravitation or acceleration. (For instance, in a freely falling elevator, one feels no gravity.) This is another way of stating Einstein's equivalence principle.

It follows that all the effects of gravitation in any frame of reference, such as one at rest on the surface of the Earth, can be described by specifying the transformation of space and time coordinates to a suitably accelerated frame, known as an inertial frame, such as an elevator accelerating downward at 32 feet/second per second, in which inertial and gravitational forces cancel.

This is how the geometry of space-time gets into the description of gravitation. It parallels the description of curved surfaces worked out by mathematicians in the nineteenth century. Everyone knows that it is not possible to make a map of the whole Earth that faithfully reproduces straight lines and areas – this is obviously because the Earth's surface is curved. But it is possible to find a transformation from latitude and longitude to other coordinates such as distance uptown and crosstown, which provide a good enough map for a small area such as a city. The transformation to an inertial frame like a freely falling elevator at any point in space and time is like the projections of maps in a geographic atlas to local coordinates like uptown and crosstown, in which in a small region the curvature of the Earth has no effect on the map.

All this has been well known since Einstein introduced the general theory of relativity, and I was probably not alone in giving priority to the equivalence principle over geometry in grounding this theory. In any case, I did go on teaching general relativity from time to time at Berkeley and then at MIT, and this was the point of view that informed my lectures and would inform my first book, *Gravitation and Cosmology*, published in 1972.

This point of view may leave one with the following question: Why does gravitation obey the equivalence principle? I will return to this question shortly.

Louise was pregnant and I did not want to leave her alone, so I did not join a JASON visit to an aboveground test of a nuclear weapon in the Nevada desert. Nuclear testing in Earth's atmosphere was given up in October 1963, by agreement among the USA, Britain, and Russia. No one since then has had a chance to see a nuclear explosion, and a good thing too.

Spending time with Louise at home, I managed to get some work done, of which I am rather proud, although it made no great splash. Partly as an offshoot of my efforts to solve strong interaction problems by introducing quasiparticles, I began to study a branch of mathematics known as functional analysis.

Among other things, functional analysis allows calculations of scattering processes to be put on a firm basis. I learned that the reason that electronic computers can calculate the probability of scattering of one particle by another in various directions without needing any tricks like my quasiparticle approach is that the equations one must solve have a property known as complete continuity. This property has a formal mathematical definition, but its significance is that it can be approximated to any desired degree of precision by simple algebraic equations easily solved by computer. I realized that the reason that physicists had always had trouble in solving problems involving the interaction of three or more particles is that the equations are not completely continuous. I saw how to rearrange these equations for any number of interacting particles so that they would be completely continuous, and therefore in principle reducible to simple algebraic equations suitable for a computer.

Then, alas, I learned that a Russian mathematical physicist, Ludwig Faddeev, had already done this for the three-particle case. I had gone beyond his work by dealing with any number of particles, so my work was still worth publishing, but the key ideas were already there in Faddeev's papers. It was a repetition of my experience in 1955, when I was scooped in my work with the Princeton Atomic Beam Group.

The real moral of the story is that particle physicists have willy-nilly locked arms, are joined together in each other's work, and, building on each other's work, we make our occasional advances. (The film *The Red Shoes* had been like this for me.) We are engaged in a worldwide, historic joint enterprise. As I write this, in 2021, the age of accelerators is nearing its end for us, but our age of cosmology is just beginning.

That summer I had begun to rethink what I had learned about the quantum theory of fields. At that time, physicists generally followed an approach that had been laid out in 1930 in a series of papers by Werner Heisenberg and Wolfgang Pauli. For each kind of elementary particle, one introduces some sort of field, a quantity that, like the electric and magnetic fields, depends on the particle's position in time and space. The physical properties of these fields are specified by introducing an action, a numerical quantity whose value depends on the values of all fields at all times and spatial positions. From the form of the action, one derives field properties using certain rules that make up what is known as the canonical formalism. For a given action, these rules dictate the field equations, which constrain how the fields change with time, and the fields' quantum properties, which dictate how the fields act on physical states. It turns out that the fields create and destroy particles.

These particles will have spin. In classical mechanics, spin, also known as angular momentum, is a measure of how much matter is rotating and how fast it is rotating. In quantum mechanics, spin is better described by the states of the particle, as they appear to change when an observer rotates the frame of reference from which the particle is observed. In the natural units that are convenient in physics, photons, the particles of light created and destroyed by the quantum electric and magnetic fields, have spin one. Electrons have spin $½$. The spin of the particles that can be created and destroyed by a given field are governed by the nature of the field – specifically, by how the field depends on both the orientation and velocity of the frame of reference used by an observer.

About this, there had been a great deal of confusion, which had not entirely dissipated by the early 1960s. Paul Dirac had proposed in 1928 that elementary particles other than the photon could only have spin $½$. This was because he thought that the way to make quantum mechanical theories consistent with the special theory of relativity was to introduce a relativistic wave function, a numerical quantity that is dependent on the time and the position of each particle, like the wave functions used in the 1925–26 development of quantum mechanics by Erwin Schrödinger. But this was a quantity whose dependence on time and position would obey wave equations that – unlike the Schrödinger wave equation – would be consistent with special relativity. That is, the equations would take the same form whatever the velocity of whoever is observing the system. Dirac thought of these wave functions as probability amplitudes in the same sense as Schrödinger wave functions: The square of the wave function at a given position would give the probability per volume of finding the particle in a small volume around that position. For this to be possible, the total probability of the particle being anywhere would always have to add up to 100 percent. Dirac found that this was only possible if the particles had spin ½.

At first this seemed like a triumph, because in 1928 the only known particles that seemed to be elementary were the electron and proton, both of which do have spin ½. But as it turned out, Dirac was wrong about how to reconcile quantum mechanics and relativity. This was already seen when physicists tried to apply Dirac's approach to problems involving several electrons. The wave function would depend on several positions, one for each electron. But does relativity, which treats space and time in parallel, then require that the wave function also depend on several different times, one different time for each electron?

Also, by now, we know that there are several types of massive particles that seem every bit as elementary as the electron, but do not have spin ½: The W and Z particles that transmit the weak nuclear force have spin one, and the Higgs boson has spin zero. Nevertheless, textbooks continued well into the 1960s to present the Dirac theory as a serious approach to relativistic quantum mechanics. It would not surprise me if they still do.

The only satisfactory relativistic quantum theories are quantum field theories. The equations of the Dirac theory describe one kind of field for particles of spin ½, but it is a quantum field, not a probability amplitude like a Schrödinger wave function. This quantum field theory leads, among other things, to the same successful results as the Dirac theory, including a very good approximation to the magnetic field of the electron.

Our equations are often smarter than we are; they may survive even when the original motivation for the equations has been invalidated.

It is possible to invent fields that can create and destroy particles of any given spin. But confusingly there was more than one possible field type for each spin. Worse, Schwinger had pointed to difficulties in the application of the canonical formalism to any field theories of particles with spin greater than one.

In the work that I began in the summer of 1963, I decided to jettison the whole Heisenberg–Pauli canonical formalism. After all, why is it required in quantum physics? The canonical formalism was originally introduced into quantum mechanics by way of analogy with methods used in the nineteenth century to study problems of classical mechanics, from the motion of planets to the properties of gases. Analogy is often a valuable source of inspiration in theoretical physics, but it never provides a legitimate explanation of why theories apply to the real world.

Instead of the canonical formalism for field theory, I would take particles, not fields, as my starting point. Without using field theory, Wigner had shown in 1939 that there is a unique description of any one-particle state in relativistic quantum mechanics, depending only on the particle's mass and spin. I could see how to construct any sort of quantum field out of operators that, acting on a physical state, can destroy or create a single particle of given mass and spin in that state. There were "minimal" fields, from which any other field for a given mass and spin could be derived as suitable rates of change of the minimal field with time and position. I saw how to introduce interactions among these fields, so that the rates for any sort of reaction among the particles would satisfy the rules of special relativity. The rules for calculating these rates followed the lead of Feynman's rules for calculations in quantum electrodynamics, so when in October 1963 I submitted the paper reporting these results to the *Physical Review*, I gave it the title "Feynman Rules for Any Spin."

This work improved my understanding of antimatter. In 1928, Dirac had discovered that his wave equation had solutions of both positive and negative energies. To explain why electrons do not all fall from positive to negative energy states, radiating away the energy difference, Dirac supposed that under ordinary circumstances the negative energy states are all filled. Pauli in 1925 had introduced an exclusion principle, according to which only one electron can occupy any given state. So if all states of negative energy are filled, no electron of positive energy could fall into any of them. (This is analogous to the reason that the electrons in atoms do not all fall into orbits of the lowest energy.)

Dirac also suggested that occasionally an electron would be knocked out of one of these negative energy states, leaving a hole that would appear as a particle of positive energy and positive electric charge, opposite in sign to the charge of the electron. Dirac's theory seemed to gain great support in 1932 when just such positively charged particles, now called positrons, were discovered in cosmic rays. But Dirac's negative energy states and holes were imaginary all the same. As a counter to Dirac, in 1934, Pauli and Victor Weisskopf introduced a quantum field theory of charged particles of spin zero, which turned out in this theory to have antiparticles of the same mass and opposite charge, even though particles of spin zero do not satisfy the Pauli exclusion principle and therefore cannot form a stable sea of negative energy states.

The spinless particles of Pauli and Weisskopf were hypothetical, but by the 1960s there were many known charged particles of zero spin, such as pions and kaons, which were observed to have antiparticles of opposite electric charge. In a conversation during a coffee break at a physics conference in Coral Gables in the 1970s, I asked Dirac why he did not think that the existence of these spinless antiparticles cast doubt on his theory of antimatter. He answered that he did not think that these particles were "important." That seemed to me a silly answer, but later I saw that he must have meant that he did not think these particles were elementary. For instance, a positively charged pion might be the composite of a proton and an antineutron, both of which have spin ½, while its antiparticle would be the composite of a neutron and an antiproton.

Fair enough, but as I have already mentioned, the $W+$ and $W-$ particles found experimentally in 1984 seem just as elementary as the electron. These particles are antiparticles of each other, but W particles have spin one and do not satisfy the Pauli exclusion principle.

In my 1963 work, I showed how relativity requires that, in a field for a charged particle of any spin, the operator that destroys a particle must be accompanied with an operator that creates an antiparticle of the same spin and mass but opposite charge. The existence of antimatter has nothing to do with spin ½ or with the specifics of the Dirac theory.

The confusion about this is illustrated by one of the arguments used by the physicists at Berkeley in the 1950s to get funding to build the Bevatron. They pointed out that this accelerator would reach energies at which, for the first time, it would be possible to create the antiproton, the antiparticle of the familiar proton. To make this seem more exciting they argued that, since the strength of the proton's magnetic field was known to be different from the strength predicted by the Dirac theory, protons were not described by the Dirac theory. Thus, it was an open question whether protons had antiparticles. It was nonsense, but at least the Bevatron was built, and did produce antiprotons.

On New Year's Eve, there was a party at the Rosenfelds' magnificent house. The high point of the party was the performance of a new song. Each stanza described one of the physicists present at the party, and ended with the line "Some day he'll win a Nobel Prize." As far as I can remember, the stanza that described me went:

> Weinberg writes a paper every week,
> Some are true and some are lies,
> Dah de dah de dah de dah,
> Some day he'll win a Nobel Prize.

I remarked to someone at the party that what bothered me was not that I had not won a Nobel Prize, but that it was not an injustice.

In late 1963, I began work on a series of papers about massless particles. Over the years, massless particles have become rather important to me in my own work. (As I write this, I have to say that my latest paper, written in 2020, was about the quantum field theory of massless particles in space-times of any dimensionality.)

This line of work began with my earlier paper, "Feynman Rules for Any Spin." I was extending that work to particles of zero mass. At first sight, it seems that dealing with zero mass would pose no problem. As long as we stick to minimal fields, we can find the rules for calculations of the rates of processes involving massless particles by just taking the rules for massive particles and letting the mass go to zero.

But the exchange of a massless particle described by a minimal field would not produce the kind of long-range forces familiar in electromagnetism and gravitation, forces that decrease only as the inverse square of the distance. And if we try to describe such forces by starting with non-minimal fields for massive particles, and letting the mass go to zero, we find results with a zero mass in various denominators, which makes no sense.

To describe long-range inverse-square law forces, it becomes necessary to use potential fields, which have rates of change with time and position that equal the minimal fields. Constructing these potential fields from the operators that create and destroy spinning massless particles, you find fields that do not transform simply when the frame of reference in which we observe phenomena is rotated, like the various fields of massive particles, but instead also undergo a shift, known as a gauge transformation.

Gauge transformations were nothing new in physics. In Maxwell's theory of electromagnetism, the electric and magnetic fields can be expressed in terms of potentials. Two potentials that only differ by a gauge transformation give the same electric and magnetic fields, and indeed give the same results for all possible observations – provided that electric charge is conserved. That is, provided that the total electric charge in any closed system does not change with time. This is known as gauge invariance.

But now I was turning the argument on its head. Because of the peculiar properties of massless particles, as described in 1939 by Wigner, any field theory of massless particles that allows for long-range inverse-square law forces – if consistent with relativity – must necessarily possess gauge invariance. This implies that electric charge, the source of the long-range force, must be conserved.

Likewise, the source of long-range gravitational forces must be conserved. This is only possible if the effect of long-range gravitational fields on any particle is proportional to combinations of the energy and momentum of that particle, with the same combinations for every particle. Since the same is true of the inertial forces produced by acceleration, this amounts to an explanation of Einstein's principle of the equivalence of gravitation and inertia.

This work involved a calculation of the rate of emission of low-energy massless particles in any reaction among other particles, which becomes exact in the limit of vanishing massless particle energy. This work has had legs. It inspired some of my own work the following year, and recently has been put to good use by others.

In other papers, I was able to show that electric charge conservation and the equivalence principle can be derived without any use of quantum field theory, just from the general principles of relativity and quantum mechanics applied to massless particles of spins one and two, known respectively as the photon and graviton. Of course the same results were long known as consequences of Maxwell's theory of electromagnetism and Einstein's theory of gravitation. But the particular equations of these theories were derived from observation, combined of course with a good deal of physical insight. Now I could see that some of the main results of these theories could be obtained without relying on Maxwell's and Einstein's equations, just from relativity and quantum mechanics and the assumed existence of certain massless particles. In the following year, I was even able to show on this basis that any quantum field theory of these particles must satisfy the equations of Maxwell and Einstein. I emphasized this point of view also in a talk I gave in June 1968 in Trieste, the first time that I visited Salam's new International Centre for Theoretical Physics.

So which is truly fundamental? The theories of Maxwell and Einstein, from which we can infer the existence of massless particles of spins one and two, or the existence of these massless particles, from which we can derive Maxell's and Einstein's theories? We do not know. In today's string theories, photons and gravitons are more fundamental; they emerge as modes of oscillations of a string. But these string theories are still very speculative, so the issue is not settled.

This is the sort of puzzle we often confront in physics: Truth A follows from truth B, and B follows from A. So which is more fundamental? By one truth being "more fundamental" than another, I think we can only mean that it will someday be derived by a more direct chain of inference from the final laws of physics, which of course we do not yet know. So these are not idle questions – their answers reflect different guesses about the nature of these final laws.

Early in 1964, I received an invitation from Stanley Deser to lecture at that year's summer school in theoretical physics at Brandeis University. I accepted and prepared two separate lecture courses. One was an overview of everything I had learned in the course of pursuing the quasiparticle idea. It was a farewell to that work, which, alas, had not led to anything very useful aside from my education in functional analysis. The other course would be on the quantum theory of massless particles, which was occupying me more and more.

I also had an invitation to the "Rochester" Conference on high-energy nuclear physics, which was to be held later in the summer of 1964 at the Joint Institute for Nuclear Research of the Soviet Union. That was pretty exciting, even apart from physics. Russia was, as it remains, the mysterious adversary country, not a place normally visited by American tourists.

Since JASON had arranged for me to have a top-secret security clearance, I was visited by a representative of the CIA, who warned me to be careful about what I said – even if I thought I was not being overheard. He also asked me to report back to him after the trip if I had heard anything that might be useful to US intelligence. I told him that I could not agree to that, because I was traveling with a wife and daughter, and if necessary I wanted to be able to say truthfully that I had no connection with US intelligence.

I do not remember the talks at the conference. For me the most important thing there happened during a break between talks. I was sitting on a stone bench, looking at the Volga, which in Dubna was a narrow stream flowing through the Institute grounds. A fellow with a straggly moustache sat down on the bench next to me and said, "Hello. I am Coleman." I think that I wittily replied, "I am Weinberg."

I knew that a Sidney Coleman had collaborated with Shelly Glashow and had a good reputation. I learned afterward when we were both on the faculty at Harvard to appreciate Sidney as having an unusually deep understanding of modern physics. His lectures were famous. I urged Sidney to write a treatise on quantum field theory, but he said that he found writing much harder than lecturing. It was only after it had become clear that Sidney would not write this book that I had a go at it.

I also learned to appreciate Sidney's sense of humor. Most of the jokes I know I learned from him, including the frozen parrot joke and the leprechaun nun joke. (Don't ask.) It was Sidney who said, "If we make any progress in physics it is because we stand on the shoulders of dwarves." He could also improvise wonderfully. Once at Harvard when he was giving a physics seminar, he saw that Shelly Glashow has fallen asleep. He stopped his lecture and, pointing to Shelly, said "This is very interesting. It is stage three sleep. Indistinguishable from clinical death." The room exploded in laughter, Shelly began to wake

FIGURE 9.1 Sidney Coleman

up, and Sidney continued, "In all but three states you could take out his heart."

Sidney was notorious for working late hours at night, and then waking up late in the morning. Once when he was asked to teach a course at 9:00 AM, he replied that it would not be possible, because he did not like to stay up that late.

We once went on an eight-mile hike with Sidney. He did such hikes regularly, for his diabetes. We both loved Sidney and grieve for his early death.

Back in Berkeley, almost immediately I put to good use my work on massless particles. A suggestion had been made in two independent papers – one by John Bell and J. K. Perring, and the other by my friends Jeremy Bernstein, Nicola Cabibbo, and T. D. Lee – that there might exist a "hyperphoton," a kind of particle with spin one, similar to the massless photon of quantum electrodynamics, but interacting with a quantity known as "hypercharge" instead of ordinary electric charge. (The suggestion was that a hyperphoton field was causing recently discovered effects that had been attributed to a breakdown of matter–antimatter symmetry.) Hypercharge was already well known to particle physicists as a quantity, like electric charge, that is conserved in strong and electromagnetic interactions. However, unlike electric charge, it is not conserved in weak nuclear interactions. It had been introduced independently in the 1950s by Murray Gell-Mann, Kazuhiko Nishijima, and Abraham Pais to explain why some newly discovered particles could be copiously produced in pairs but decayed only slowly.

The idea was that particles with opposite nonzero values for the hypercharge could be produced in pairs, without violating hypercharge conservation. But hypercharge could not be conserved in their decay into particles of different hypercharge or no hypercharge, so these decays would have to involve the weak interactions and would therefore be relatively slow. Because hypercharge, unlike ordinary charge, is not exactly conserved, the hyperphoton, unlike the photon, would have to have a small mass. But from my earlier work, I knew that there would be trouble in dealing with nearly massless particles of unit spin, if the quantity with which they interacted were not conserved. I showed that, in any of the known decay processes in which hypercharge is not conserved, hyperphotons would have been emitted so copiously that they would already have been detected, and since they had not been detected, it could be concluded that they did not exist.

I happened to be visiting Cal Tech when theorist and fellow JASON member Fred Zachariasen gave a survey talk on recent developments in particle physics. Fred mentioned my demonstration that hyperphotons do not exist. Feynman was sitting near me in the audience and leaned over, remarking, "That was clever. How did you think of it?" It was disparaging – not a compliment.

I appreciated that Cal Tech and Berkeley had a certain unity; we were always attending each other's talks, to our mutual benefit. But Louise told me I always seemed on edge going down to visit Cal Tech. I confessed to her that Murray and Dick were not very friendly to me. I think it was only partly that I was brash and self-confident. Rather, I was too often thinking outside the box, and it was their box.

Around then I began for the first time to do some nontechnical writing. Physicists working on elementary particles were increasingly eager to push beyond the energies available at Berkeley and other existing laboratories, and build the next generation of higher-energy accelerators. Ironically, the efforts of physicists at Berkeley and elsewhere to get funding for a larger new accelerator were successful in 1967, but the planned new accelerator would be too large to fit in the Berkeley hills. Instead, it would be housed in a new laboratory on the Illinois prairie, later named Fermilab. Berkeley would never again be the leading center of experimentation on elementary particles.

With the war in Vietnam, politics at Berkeley became divisive. Years later, I would find that it was much easier to be a liberal in Texas than a conservative in Berkeley. I write this in 2020, in the genial, tolerant, welcoming atmosphere of Austin, which Louise appreciated back in 1979 when she first visited and gave a course there. It was part of what attracted me when I decided to try to join her at the University of Texas at Austin two years later and took up my present position the following year.

The 1965 JASON summer study was held at Otis Air Force Base on Cape Cod. We stayed nearby in Falmouth. The house we rented in Falmouth had a television set in the study. I believe that was where I got into my lifelong habit of leaving the TV on while I worked.

The years at Berkeley were not the most productive I have had. But I had done some outside reading, among the great historians and prose stylists. Until my final spring in California, I had not done anything to which other physicists absolutely had to pay attention. But those may have been my most contented years. I received salary increases at faster than the standard rate, and in 1964 was promoted to full professor.

At about this time, Louise made up her mind to apply to law school and was accepted by every law school to which she applied. Louise decided to enter the Harvard Law School in the autumn of 1966.

I was not unhappy to be moving for a while to Cambridge, and the Berkeley physics department kindly gave me a leave of absence. Viki Weisskopf invited me to visit MIT, but I thought that since Louise would be at Harvard it would be more convenient for me to be there too. I still had a year left on the Sloan Fellowship that had supported us in London, so I could pay my own salary and could take a break from teaching. Harvard at first offered me the position of Research Associate, which seemed infra dig, but when I demurred they sweetened the deal and offered me a distinguished title as Loeb Lecturer. I would only have to give a few lectures on my own research during the year.

Before accepting, I wanted to make sure that this visit would be welcome to the senior theorist at Harvard, Julian Schwinger, who was in Sweden picking up his Nobel Prize. I telephoned him at the Grand Hotel in Stockholm, where I knew that laureates were housed, and reached him in the hotel lobby just as he was checking in. He was a little surprised by the call. I asked him whether I had his approval for my Loeb Lectureship. He said it was fine with him. So that was settled. I had earlier been invited to spend 1966–67 as a visiting fellow at Clare College, Cambridge, so for me the finality of our decision to go to the American Cambridge was underlined when I wrote to the master of Clare College to tell him that I would not be able to accept.

During the 1965 Christmas break, we flew to Boston to find a house to rent in our first year in Cambridge. At a party at Julian and Clarisse Schwinger's house, we heard that Bernard Feld and his family would be away in 1966–67, and were looking for someone to rent their house. It was a large house in the pleasant neighborhood of old Cambridge, to the west of the Harvard campus and next to Viki Weisskopf's house, so we were delighted at the chance to rent it.

Early in 1966, I wrote my first book review. The *New York Times* asked me to review a collection of essays dealing with various aspects of time. I made the mistake of commenting on most of the essays, which made my review long, disorganized, and boring. The *New York Times* has never again asked me to review anything. The one thing about the review that I remember with pleasure was my attack on essays by Olivier Costa de Beauregard and by Herbert Dingle. Those writers

FIGURE 9.2 Julian Schwinger, c. 1964

philosophized about the role of time in physics. A French philosopher wrote to me afterward to say that as a North American I could not be expected to understand philosophical reasoning.

In the spring of 1966, I started work on a topic that would occupy me fruitfully for years to come. At the time, it was known as "current algebra." I touched on this in Chapter 7. Weak interactions like those responsible for nuclear beta decay are generated by several currents, which radiate electrons and neutrinos in something like the way that electric currents radiate light. These currents are operators; that is, one of these currents acting on a nuclear state changes it to another state. When one current acts after another, the result depends on the order in which they act. This dependence on order is prescribed by what is called the current algebra.

The conservation of currents follows from the existence of a symmetry of the equations of the underlying theory. The details of the current algebra depend on just what that symmetry is. All this holds, even if the symmetry is broken – that is, even if the solutions of the equations of the theory do not share the symmetry of those equations.

In 1965, a dramatic use was made of the algebra of the weak currents by two young theorists, Stephen Adler at Harvard and William Weisberger at Stanford, working independently. They were able to calculate the strength with which one of the currents, the axial vector current, acts on proton or neutron states, finding a result in agreement with beta decay experiments. Suddenly current algebra became a hot topic.

In the spring of 1966, I had an idea about other uses of current algebra. The assumed algebra and approximate conservation of the vector and axial vector currents pointed to the existence of a broken symmetry of the strong interactions, a symmetry under rotations of proton and neutron states into one another, but acting differently on particles spinning to the left or right of their direction of motion. That seemed to me to be pretty interesting in itself, apart from any application to beta decay.

Nambu and his collaborators in 1962 had used the approximate conservation of the currents to derive formulas for the rate of emission of a single relatively low-energy pion in a reaction among other higher-energy particles. For this purpose, they did not need to know anything about the current algebra, which depends on the details of the symmetry underlying the conservation of the currents. I undertook to calculate the rates of processes involving more than one low-energy pion, which I could see would depend on both current conservation and current algebra – that is, on the detailed underlying symmetry.

The first problem I attacked proved too difficult. Recalling my own calculation in 1965 of the rates of processes involving any number of low-energy photons and gravitons, I tried to do the same for low-energy pions. Because pions can be radiated from pions, this calculation was beyond me, and so far has been beyond anyone else.

Then I turned to a problem that was easier and also more important, to calculate the rate at which a low-energy pion is scattered by a proton or neutron into a pion traveling in another direction. I was able to get an answer by taking the current algebra to be the same as assumed by Adler and Weisberger. (The same calculation was done independently by Y. Tomozawa.) The result agreed with existing experimental values. I was even able to use this result to give an alternative derivation of the Adler–Weisberger result.

I also considered a harder problem, the scattering of a low-energy pion by a pion. The result here depended not only on the current algebra but also on the mechanism responsible for the approximate nature of the symmetry and the resulting nonzero pion mass. Making the simplest possible assumption about this mechanism, I was able to get answers for the scattering rates for the different pions with different charges.

These results made a stir. Pion scattering was just the sort of problem that the followers of S-matrix theory had tried to solve. The progress that they hoped to make would depend on the assumption that this scattering is strongest at low energy. I had found definite results without any such approximation, and these results indicated that pion scattering is weakest at low energy. Specifically, the effective diameters of any target particle in the scattering of low-energy pions (known as the scattering lengths) are proportional to pion mass, by far the smallest mass of any strong interacting particle.

During that summer, I was too much wrapped up in this work to participate full time in the JASON summer study. The 1966 summer study was held in Santa Barbara, one of the loveliest towns in California. Instead of taking a house there for the summer, we made a few flying visits, staying in a motel. I was working on the instabilities of a beam of charged particles, work later published in the *Journal of Mathematical Physics*. As often happened in my work for JASON, the experience was educational. A good deal of what I learned about the theory of instability proved useful years later when I had to teach about the instability of stars.

The 1966 "Rochester" Conference on high-energy nuclear physics was held at Berkeley at the end of the summer. I was tapped to be chairman of the session on current algebra, a position of no importance, but sitting on the stage I had a rather gratifying experience. My work on pion scattering was described in some detail by the rapporteur, Roger Dashen.

A few days before we were due to fly to Boston, our dear friends, Elaine and Ken Watson, had us over to their house in Berkeley for dinner with Murph and Mildred Goldberger, and goodbyes.

10 Cambridge: 1966–69

The work that led me to develop my ideas about effective field theory began in a place closer to our own house, as I sat at the counter of Brigham's, a coffee shop in Harvard Square (alas, one of the landmarks now gone), doodling on a napkin.

I had thus far failed to say anything useful about the rate of emission of low-energy pions in reactions among other higher-energy particles. Using current algebra to write down these rates for just two or three emitted pions, it struck me that the results looked very much like what one would calculate using Feynman diagrams in a quantum field theory. That was odd. I realized that any quantum field theory that had the right symmetry properties would entail the existence of currents that satisfy the conditions that are assumed in using current algebra, so if such a theory led to definite predictions for any process, these predictions would have to agree with the predictions of current algebra. I knew of a theory with the right symmetry properties, known as the sigma model, studied a few years earlier by Bernstein, Fubini, Gell-Mann, and Thirring, and by Gell-Mann and Lévy. But this was a theory of strong interactions, not a theory like quantum electrodynamics, in which, because the contributions of complicated intermediate states are suppressed by powers of the small number 1/137, valid approximate results could be calculated, including just a few intermediate states. Indeed, the whole point of current algebra had been to derive results for processes involving strong interactions, where the methods of calculation used in quantum electrodynamics would not work.

Then I remembered that, in the calculations I had done at Berkeley, calculations of the scattering of low-energy pions on various targets, the scattering rates turned out to be proportional to the square of the pion energy, which is why they were unexpectedly small. I saw how to rewrite the sigma model so that the simplest contributions to low-energy pion scattering would be proportional to the squared pion energy, and more complicated terms would be suppressed by higher powers of the energy. This rewritten theory is one that would make definite predictions in the limit of low pion energy. And, since it satisfied the assumptions made in current algebra, its predictions had to agree with the predictions of current algebra – except that they were much easier to derive.

Current algebra was still lurking in the background, as the rationale for these field theory calculations. Neither I nor anyone else at the time took the rewritten sigma model seriously as a basis for calculations of any but the lowest approximations. The field theory was just a convenient tool. That would change with the advent of modern effective field theory, but that change was more than a decade away. Beautiful!

That winter I concentrated for a while on current algebra. I followed pretty much the same path that Adler and Weisberger had followed in their successful calculation mentioned in Chapter 9. But this time, I would work out the consequences of current algebra when the currents of the weak interactions act on the vacuum state instead of the one-nucleon state. The results were a set of exact formulas that I called spectral function sum rules.

By themselves, these formulas made no predictions that could be compared with observation. But with some plausible further assumptions, I was able to derive a striking result: I saw that the well-known strongly interacting particle of spin one, known as the rho meson, which decays into two pions, had to have a sort of partner, a strongly interacting particle of spin one that decays into three pions, with a mass equal to the rho meson mass times the square root of two. Just such a particle called "the a_1 meson" was already known, so this seemed like a great success. I didn't quite trust the approximations I was making and would come back to this calculation the following autumn, with fateful results.

I had an offer of a professorship at MIT, which was especially attractive to me because of the presence there of Francis Low, whose work I found inspiring. Francis and Natalie Low would become increasingly treasured friends.

JASON's meeting that summer was the most contentious such meeting that I would ever attend. Feelings about the Vietnam War were heating up, and were affecting the work of the group. Some members, led by oceanographer Bill Nierenberg, Director at Scripps Institution of Oceanography in San Diego, did what they could to help the war effort, chiefly by trying to design methods to block the Ho Chi Minh trail that was being used by the North Vietnamese to send supplies and troops to the south. I can believe that Bill and others thought this was the way to end the war most quickly. But others, including myself, wanted nothing to do with the Vietnam War at all. I planned to devote myself to more traditional JASON tasks having to do with the strategic balance with the Soviets.

Then we heard a rumor that powerful figures in the Pentagon wanted to introduce tactical nuclear weapons in Vietnam. Several of us regarded this as morally repugnant and strategically disastrous. We also thought it was stupid. The Viet Cong forces were spread out and hard to locate, so our use of nuclear weapons would be ineffective against them, while the US had large fixed facilities that would make good targets for nuclear weapons that might be furnished to the North Vietnamese by the Soviets.

JASON members Freeman Dyson, Bob Gomer, Courtenay Wright, and I decided to do a study of tactical nuclear weapons in Vietnam that might help to prevent their use. We didn't think that anyone would be impressed by our credentials for making moral or strategic judgments, so this would be a cold-blooded technical study of the military implications of the introduction of nuclear weapons in the Vietnam War. Not surprisingly, we concluded that it was a dumb idea. The analysis was done honestly, but I have to admit that if I had not expected this result I would not have participated in the study.

Our analysis was presented in a secret report with the title "Tactical Nuclear Weapons in Southeast Asia." Several people, including the Institute of Defense Analysis consultant Seymour Deitchman and JASON geophysicist Gordon MacDonald, have been quoted as concluding that our report was influential in putting an end to the issue. I have no way of judging this, but apparently the rumors of plans for using nuclear weapons in Vietnam had been accurate. On October 7, 2018, the *New York Times* reported that fifty years earlier General William Westmoreland, the American military commander in Saigon, had planned to move nuclear weapons into South Vietnam, but had been overruled by President Johnson.

Unfortunately, although our JASON report remained classified for many years, the title appeared in unclassified lists of JASON studies, giving a misleading impression of the authors' views. A few years after that 1967 summer study, on a visit to San Francisco, Louise and I drove over to Berkeley to take a nostalgic look at our former home, and found a chalked message on the pavement outside the house: "Steven Weinberg, War Criminal." The 1967 report was finally declassified as a result of a Freedom of Information Act request, and can now be downloaded by typing its title in a Google search bar, so anyone can see what we were trying to say.

During evenings and weekends that summer, I continued my work on the use of field theory to reproduce the results of current algebra, leading me further toward the idea of effective field theory. In my previous work, beginning at the counter of Brigham's coffee shop, I had started with a field theory, the sigma model, which had the symmetry properties that underlie current algebra. To be specific, there are four fields in this model that behave like the projections of a vector in a mythical four-dimensional space along four perpendicular directions. I had rewritten this theory by expressing these four fields in terms of four other fields: the length of the vector, and three angles that specify the direction of the vector. I then threw away the field, defined as the length of the vector, which plays no role in the symmetry properties of the theory, so that the theory just had three fields, which I identified as the fields of the three types of pion that are distinguished by their electric charges, respectively equal to the electron charge, to minus the electron charge, and to zero.

Late in the spring of 1967, Julian Schwinger had suggested to me that I did not need the sigma model; I could just directly work out the structure of a theory involving three pion fields with the symmetry properties of these three angles. After all, why should anyone take the sigma model seriously as anything but one illustration of a broken symmetry? I was not proficient in using group theory, the mathematical theory of symmetry, so I spent a good deal of time that summer working out this structure by a sort of brute force. Finally, I was able to recover the previous results I had found from the sigma model, now with less in the way of unnecessary assumptions. The following autumn this work was generalized to arbitrary symmetries with arbitrary patterns of symmetry breaking by a quartet of theorists more learned than I in group theory: Curt Callen, Sidney Coleman, Julius Wess, and Bruno Zumino.

I arrived at MIT in September 1967, and became an instant hero. The Center for Theoretical Physics was just then moving into handsome new quarters on the third floor of one of the mazes of connected buildings that house most MIT faculty offices. Under the influence of Viki Weisskopf, the Center was designed to promote spontaneous conversations among theorists. Its offices were connected by a large carpeted hall, equipped with blackboards, benches, and some handsome modern paintings. I heard that, to maximize the efficiency of the air-conditioning system, all office windows would be sealed. I let it be known that, if my office window were sealed, I would throw my desk chair through it. The windows were not sealed. Years later, I returned to MIT for a celebration of the founding of its Center for Theoretical Physics, and heard a talk that referred to this episode as my finest hour.

That autumn of 1967, I did the most cited work of my life. It began with my return to the problems raised by my work the previous spring on spectral function sum rules. Not trusting my earlier

FIGURE 10.1 The author while a professor at MIT

prediction that the a_1 and rho meson masses would have the ratio 2, I thought of checking whether the same result would be found in a quantum theory of rho and a_1 fields. For these purposes, I adapted a class of field theories introduced in 1954 by C. N. Yang and R. L. Mills to the particular symmetry underlying current algebra. The Yang–Mills theories had been pretty well abandoned because they had predicted that the particles described by their fields would have zero mass, like the photon of quantum electrodynamics, and of course such particles were not observed. Pauli had been quite scornful about this. But now I was dealing with a broken symmetry, and I found that, in this theory, the a_1 particle was massive and heavier than the rho particle, just as expected from the spectral function sum rules. But, alas, the mass ratio was not 2. Instead, the rho particle had zero mass, in violent disagreement with observation.

I should have anticipated all this. Several groups of theorists had pointed out an exception to the theorem about broken symmetry that had been published two years earlier by Goldstone, Salam, and Weinberg (me), which had dictated that a massless spin-zero particle would appear for each broken symmetry. As mentioned in Chapter 9,

Brout and Englert; Higgs; and Guralnik, Hagen, and Kibble had shown that, in theories with spin-one fields of the Yang–Mills type, there would not be a massless particle for each broken symmetry. Instead, what would have been a massless particle of spin zero would be one of the states of the spin-one field that interacts with the current of that broken symmetry, giving the quanta of that field a finite mass. That is why I was finding that the a_1 meson was massive.

Kibble had further emphasized that, if any part of the symmetry were not spontaneously broken, then the corresponding spin-one particle would remain massless. That is why I was finding a massless rho meson. It is associated with the isotopic spin symmetry between protons and neutrons, which remains unbroken in the breakdown of the larger symmetry group of which it is a part.

So far, this thinking, which seemed to be getting us closer and closer to good theory, seemed to be getting us farther and farther from experimental reality. Rho mesons are not massless.

Then one day, while driving to MIT in my red Camaro, I had a good idea. This was the idea that would be the subject in 2017 of an international conference convened at Case Western University to commemorate the 50th anniversary of this work, which, putting together the thinking in so much fine earlier work, had turned out to be key to formation of the modern Standard Model of elementary particles, and would gain me the Nobel Prize. Friends and other colleagues from all over the world flew in to celebrate the occasion, and I gave the summary talk.

As I would explain at Case Western in 2017, at some point in the autumn of 1967, I realized that I was working on the wrong problem. Maybe in a theory of weak and electromagnetic interactions, of course with a different spontaneously broken local symmetry group and different matter fields, the massive spin-one particle would turn out to be not the a_1 meson, but the W particle that had long been supposed to transmit the weak force. And the massless spin-one particle would not be the rho meson, but the photon, associated with unbroken electromagnetic gauge invariance. So without expecting it, I was on my way to a unified theory of weak and electromagnetic forces.

To construct a realistic theory, I needed to assume something about the matter particles with which the photon and massive spin-one particles interact. At that time, I had no faith in any existing theory of strong interactions, so I decided to limit the matter particles in the theory to just those that have only weak and electromagnetic interactions: the electron and electron-type neutrinos and similar other particles, collectively known as leptons. Once this decision was made, the choice of symmetry group was almost inevitable. There would have to be four photon-like fields of spin one. When the symmetry was spontaneously broken, these four fields would show up as the massive $W-$ particle that had been widely supposed to transmit the weak force in nuclear beta decay – its antiparticle, the $W+$ particle; the massless photon of quantum electrodynamics; and a new electrically neutral particle heavier than the W, which I called the Z particle. I chose the name "Z" because that particle would have zero electric charge, and Z was the last letter of the alphabet, so it is peculiarly appropriate for the last member of the photon's family.

I also had to introduce some mechanism to produce spontaneous symmetry breaking, to give mass to the W and Z particles and also to the electron and to heavier charged leptons, like the muon. I chose the simplest possibility, a single multiplet containing four fields of zero spin: one positively charged, one negatively charged, and two neutral. As a result of the symmetry breaking, the charged spin-zero fields and one of the neutral ones did not manifest themselves as physical particles, but served instead to give mass to the $W+$, the $W-$, and the Z particles. The other neutral spin-zero field did show up as a physical heavy particle, which I later called the Higgs boson, as Peter Higgs had predicted a boson in connection with symmetry breaking.

On October 15, 1967, I sent an article on this theory, "A Model of Leptons," for publication in *Physical Review Letters*. It was to become the most cited article in the history of theoretical physics, and it may still be. But at the time, my article just lay there, like a wet newspaper.

This was because it was not clear that this theory had solved the problem that had bedeviled all previous theories of weak interactions, the problem of infinities. In quantum electrodynamics, all of the infinities that are encountered in calculations could be eliminated by a redefinition, or renormalization, of the charge and mass of the electron and the scale of the electromagnetic and electron fields. Theories in which infinities could be eliminated in this way are called renormalizable. The existing working theory of beta decay due to Enrico Fermi was not renormalizable, and so could not be used in any but the lowest order of approximation. After working out the model of leptons, I realized that its structure was similar to that of a theory proposed earlier by Shelly Glashow, but his theory did not introduce spontaneous symmetry breaking as a source of the W and Z masses and, in consequence, was not renormalizable either. The future of the model of leptons hinged on whether it was renormalizable. In my article, I noted that the theory with no spontaneous symmetry breaking and massless W and Z particles was probably renormalizable, so the issue was whether the renormalizability was preserved when the symmetry was broken and the W and Z became massive. I thought it was, but at that time could not prove it. A little later, Salam independently constructed the same theory, and expressed confidence that it was renormalizable, but could not prove it either. So this theory, to which Salam gave the name electroweak theory, at first attracted only a little attention.

That autumn, I took the family to Brussels, where I would attend the Solvay Conference, "Fundamental Problems in Particle Physics." This was the 14th in a historic series of famous physics conferences, going back to 1911. Attendance was select, limited to a small number of invitees. The smallness of the gathering encouraged particularly interesting exchanges during question periods, or in informal conversations.

T. D. Lee gave a talk about a theory of his that connected weak and electromagnetic interactions, in which the mass of the W particle that transmits the weak interaction was predicted to have the value (in equivalent energy units) of 37.3 billion electron volts. In the question period after his talk, I mentioned that I had what seemed like a promising theory of weak and electromagnetic interactions, in which the W particle would be heavier than 37.3 billion electron volts, with an even heavier neutral companion, heavier than 74.6 billion electron volts. I could not promise that the theory was renormalizable, so enthusiasm was tepid. But one of the participants, Hans-Peter Dürr, took away the handwritten draft of the paper that I was submitting to *Physical Review Letters*. This draft survived a long while; it surfaced half a century later in a photocopy sent me by the physicist-historian Frank Close.

One day at the meeting, I was standing at a wall where the sepia-toned group photos of participants at earlier convenings of these famous conferences were hung. In the 1911 photo, I noted Albert Einstein, Max Planck, Ernest Rutherford, and a single woman, Marie Curie. Suddenly I became aware of an elderly physicist with a goatee standing next to me and looking over the same pictures. Then I did a double take. The same man was there in the 1911 photo, looking hardly any younger! The labels in the picture's border indicated that the man in the photo was Jean Perrin, who in 1908 had verified Einstein's theory of diffusion and used it to give one of the first good estimates of atomic masses. Then the man standing next to me introduced himself as Francis Perrin. He was the son of Jean. (I would come to know Francis a little better a few years later, when he was my host at the Collège de France.)

These Solvay photos are famous, and are frequently used as illustrations in books about science. Collections of the photos show the participants steadily growing older (and Einstein's hair steadily getting wilder). I used the photo of the 1911 conference in my book *Gravitation and Cosmology*. So it was satisfying to me that now I too would be in one of the Solvay Conference group photos. Or so I thought.

A few days later, there was to be a talk by Werner Heisenberg. Of course I knew of him as one of the founders of modern quantum

FIGURE 10.2 The 1911 Solvay Conference in Physics

mechanics, who had invented the version known as matrix mechanics, had deduced the uncertainty principle, and was one of the first to apply quantum mechanics to fields. A truly great physicist. But I did not want to hear his talk. For one thing, I was pretty sure that Heisenberg would have nothing interesting to say. When I was a postdoc at Columbia, I had attended a famous secret seminar in which Pauli described a theory he was developing with Heisenberg. It offered no really new ideas, and it could not be used to calculate anything – not only because its interactions were strong but also because calculations using it would inevitably encounter infinities. Somewhat embarrassed by the lukewarm reception of the audience, Pauli commented to Bohr, who was sitting in the front row, that perhaps these ideas were too crazy. Bohr replied that no, they were not crazy enough. Pauli should have known better, and soon did. There is a story that he sent a friend a report on the new theory, saying that only the details needed to be filled in. It was a blank sheet of paper. But Heisenberg continued to promote this theory. I heard from a German experimental physicist that Heisenberg was opposing the development of new experimental facilities for elementary particle physics in West Germany, because the highest priority should be given to theory, specifically, his theory.

Even so, I might have attended Heisenberg's talk, if only out of politeness. But I did not want to go out of my way to be polite to Heisenberg. In Copenhagen, I had read Goudsmit's book *Alsos* (1947), which convinced me that Heisenberg had done his best to give Germany a nuclear weapon as head of its atomic bomb project in World War II. The world was spared this disaster in part because of failures of Heisenberg's judgment. Heisenberg never realized that, even if he had built his reactor and used it to make plutonium, it would produce a mixture of plutonium isotopes, and therefore could not be used to produce a nuclear explosion without being triggered by a spherical implosion of conventional explosives beyond the current state of the art in Germany. But he did his best.

As far as I know, Heisenberg was no Nazi, but he was an ardent German nationalist and devoted his talents to Nazi Germany. Viki Weisskopf once told me of a conversation he had with Heisenberg shortly before the war. Heisenberg said that he regretted the likelihood of war, with the suffering it would bring, but took comfort from the fact that it would spread German culture throughout the world.

So on the day of Heisenberg's talk, I drove out with Louise and Elizabeth to look at the field of the Battle of Waterloo. Returning later to the meeting, I found that, alas, I had missed not only Heisenberg's talk but also the group photo. When the 1967 group photo was published, the margin contained a note "Absent: S. Weinberg and M. Gell-Mann." I have no idea why Murray was not in the photograph, but I like to think that deciding not to attend a Heisenberg talk was one of the things about which Murray and I could agree.

Back in Cambridge, I returned to the problem of the renormalizability of the electroweak theory. I could see what the problem was, though not how to solve it. In applying quantum mechanics to the electromagnetic field, one must first say something about how to

FIGURE 10.3 The 1967 Solvay Conference in Physics

choose the potentials whose rates of change give the electric and magnetic fields. (This is known as a "choice of gauge," and theories of these potentials are known as "gauge theories.") It is possible to choose the potentials so that, when the theory is quantized, the potentials create and destroy photons only in their known physical states. (This is known as "unitarity gauge.") For a given photon momentum, there are just two states, distinguished by whether the photon is spinning to the right or left. The problem with this is that, although the theory is still consistent with Einstein's special theory of relativity, this is not manifest in unitarity gauge. As a result, when, for example, one calculates the effect on an electron of emitting and absorbing a photon, it is far from obvious that the infinities arising in the correction to the electron's energy and momentum are related in the way required by a relativistic theory. It is therefore not clear that these infinities can be eliminated by a redefinition of the electron's mass.

The great success of quantum electrodynamics in the late 1940s was achieved in one approach by introducing a different set of potentials, which create four rather than two kinds of photons: the two physical photons spinning to the left and right and two others that are not spinning, one of them that is created and destroyed with negative probability. Of course, negative probability is nonsense, but one can show that, in this formalism, the total probability of any physical process that conserves electric charge is always positive. The great advantage of this choice of potential is that the theory is not only relativistic; it is manifestly relativistic, so that one can see how to deal with infinities by renormalizing the electron mass and other quantities.

My problem was that I could not see how to introduce this sort of potential in gauge theories more complicated than electrodynamics, such as the electroweak theory, without risking the disaster of negative probabilities for physical processes. The year 1968 brought distractions that took me away from this problem, so I handed it over to a graduate student.

The 1968 "Rochester" Conference on high-energy physics was to be held in Vienna at the end of the summer. I was invited to give the rapporteur talk on current algebra, and spent much of my time in New York that summer preparing my lecture. This was to be the first time that I would give a plenary talk at an international conference. It was not easy. As I mentioned at the beginning of my talk, in the two years since the previous Rochester Conference, I had received 55 kilograms of preprints related to current algebra. I did my best to review all this, helped by my bright former student Lay Nam Chang, who was then a postdoc at MIT.

Along with this massive review, I decided to take the opportunity of this talk to push my point of view, which we should concentrate on the symmetries of the strong interactions that underlie current algebra, and not be so fixated on the fact that the currents arising from these symmetries happen to play a role in the weak interactions. When I gave my talk, Murray Gell-Mann, sitting in the front row, was visibly annoyed, and growled some negative remarks, which do not appear in the conference proceedings. He was, after all, the first physicist to work out the algebra formed by the currents of the weak interactions, and my perspective put his at a discount.

A fine babysitter at our hotel made it possible for us to go out in the evening several times with our dear friends Bruno and Shirley Zumino. Once when we were together in a beer cellar, Bruno, who was fluent in German, suddenly said in a hushed manner that we had to leave right away. Once outside, he explained that he had overheard remarks about Louise and me that were not only anti-Semitic, but threatening as well. Neither Louise nor I looks particularly Semitic, but Austrians may have a special expertise. Anti-Semitism is an old story in Vienna. Jews were not allowed to live there until 1848. In 1897, Karl Lueger became mayor of Vienna after forming the explicitly anti-Semitic Christian Socialist party. Hitler took Lueger as an inspiration. And yet Viki Weisskopf and other Jews like Stefan Zweig who fled Vienna in the 1930s remembered it lovingly. Viki seemed to me like a man who was hopelessly in love with a beautiful woman by whom he is endlessly betrayed. I think the hostility to Jews in Europe may be disregardful of the poverty in which many Jews live, and overly concerned with Jewish successes in finance and in the arts and sciences, successes seen as diminishing the opportunities of other Europeans.

Back in Cambridge in the autumn of 1968, I heard that I had been elected to the American Academy of Arts and Sciences. This academy was founded in 1780 by John Adams, John Hancock, and other Bostonians, perhaps as a response to the founding in Philadelphia of the American Philosophical Society in 1743. Both academies include members from a broad range of academic and artistic disciplines and from all over the United States and other countries, but the activities of the American Academy are centered in Boston. The most important activities for us were the monthly "stated meetings" then held at the Academy's handsome headquarters, Brandégee House. After dinner, there would be a talk by a member. (Bostonians love lectures. There is an old story about a Back Bay matron who, as a reward for a good life, was offered a choice between heaven and lectures about heaven, and chose the lectures.)

These meetings allowed us to make friends beyond the limited worlds of physics and law at MIT and Harvard, including Dan and Pearl Bell, Jack Fine, Nat and Lochi Glaser, Stephen and Margaret Graubard, and Mickey and Phyllis Keller.

It was through the Academy, I think, that we also came to know and love Ali and Marjorie Javan. Ali had invented the gas laser at Bell Labs.

We also found ourselves unlikely close friends, through the Academy, with Tom and Ramelle Adams. Also dear to us were other members of the Academy we knew from the world of physics: the Lows, the Jackiws, and the Frisches.

In November 1968, Richard Nixon was elected president, the greatest American political comeback since the election of Grover Cleveland in 1892. I suppose Nixon's victory was partly due to a reaction against such urban unrest as the Watts riots and the tumultuous Democratic convention in Chicago. It was partly due also to the fact that many on the left were disenchanted with Hubert Humphrey, the Democrats' candidate, for his support of the war. In any case, this election did not soothe college campuses.

The war in Vietnam was a subject of political tension, although hardly on the scale we had seen in Berkeley. In an effort to keep lines of communication open, the MIT administration decided to host a small meeting among selected students and faculty and a few government policymakers. The meeting was held early in 1969 at Endicott House, a grand house outside Boston that had been willed to MIT. At the meeting were US Senator Albert Gore Sr., future Secretary of State Cyrus Vance, and President Nixon's National Security Advisor, Henry Kissinger. There was also a young defense consultant from the RAND Corporation, Daniel Ellsberg, not yet famous for releasing the Pentagon Papers. I don't recall the names of the faculty and students present, except that I was one of them. Kissinger gave an optimistic talk about the Vietnam War, during which Ellsberg kept asking him if he knew how many Vietnamese had been killed in the war. It became clear that this was not a calculation anyone had bothered to do.

That winter I was caught up in a different public issue. Soon after the inauguration of Richard Nixon, the new Secretary of Defense, Melvin Laird, called for implementation of a program of defense against attack by nuclear-armed ballistic missiles. This program, known as Sentinel, had begun in the Johnson administration. Senator Edward Kennedy asked Secretary Laird to defer deployment of Sentinel, pending a review of strategic weapons systems.

Early in February 1969, Senator Kennedy asked Abram Chayes and Jerome Wiesner to supervise the preparation of an independent nongovernmental evaluation of the Anti-Ballistic Missile (ABM) issue. Chayes was a Professor of Law at Harvard who had been legal advisor to the State Department in the early 1960s. (He had been Louise's professor in a course on international law. His wife Toni later told us when we had become good friends that Abe had been "very impressed" with her.) Wiesner was Provost of MIT and had been Science Advisor to the President from 1961 to 1968. To help in organizing the study and to join in writing the study's "Overview" chapter, Chayes and Wiesner called in George Rathjens and me.

Rathjens had worked at the Advanced Research Projects Agency, the Weapons Evaluation Group at the Department of Defense, the Arms Control and Disarmament Agency, and the Department of State, and was currently a visiting professor at MIT. Later he was one of the founders of the Defense and Arms Control Studies Program at MIT. I was very much the junior member of this group, enlisted I suppose because Wiesner knew me and because work with JASON had given me some familiarity with missile defense technology.

Much of the debate regarding ABM had to do with cost and effectiveness, but what mostly concerned me was the effect on the arms race. The US and Soviet Union were each deterred from using nuclear weapons by the certainty that a nuclear first strike would be followed by a crushing nuclear retaliation. Even the possibility of missile defense would weaken this deterrence, impelling the other side to increase its offensive forces (or even worse, to put their missiles on a hair-trigger alert). We saw this later, when President Reagan spoiled the prospects for a large cut in offensive weapons by insisting on US development of the "Star Wars" missile defense system. The fact that it would probably not work made little difference – even the possibility of US missile defense precluded the possibility of a reduction in Soviet missile forces.

Secretary of Defense Laird had paid lip service to this consideration. In March 1969, he explained to the Senate Armed Services Committee that the Nixon administration was shifting from the Sentinel system's city defense to a new "Safeguard" system oriented toward defense of missile silos in part because "the original Sentinel plan could be misinterpreted as ... and in fact could have been ... a first step for the protection of our cities."

But he was being disingenuous. As I showed in an article in the Chayes–Wiesner book, despite the shift in the location of defense missile launching sites, the Safeguard system would give American cities pretty much the same thin defense as the Sentinel system. The real aim in changing systems was to defuse opposition from American citizens who did not want any sort of missile launching sites in their backyards. This gave me a good introduction to the flimflam that frequently attends political debates.

In 1969, missile defense was only one of the ingredients along with civil rights and Vietnam, in the boiling pot of American politics. There was a strong feeling in many universities, and certainly at MIT, that faculty members should do more to influence national policy. I was one of about forty MIT faculty members who met on March 4 (partly at the instigation of a visiting professor of physics from Cornell, Kurt Gottfried) to organize ourselves for political activity. This small gathering evolved into the Union of Concerned Scientists, a national organization that, after fifty years, is still contributing its expertise, however disregarded, to national debates on arms control, environmental policy, and much else.

In addition to this political involvement, in those days, I was also serving on a number of committees – far too many committees. I was on the Council of the American Physical Society and on several of its committees, and I served on the Physics Advisory Committee of the Cambridge Electron Accelerator (which gave me a parking spot near the Harvard Law School). Also, I was on several committees of the American Academy of Arts and Sciences.

Despite all this, I did manage to do a little work on current algebra, partly with friends like Shelly Glashow at Harvard and Howie Schnitzer at Brandeis. Even so, I began to dream of a move to a small well-regarded liberal arts college like Bowdoin, where I would be let alone to think, read, and write, and no one would ever again ask me to serve on an interesting committee.

11 Cambridge: 1969–72

Now a distinguished graduate of Harvard Law School, Louise received an offer of an associateship at one of Boston's old-established law firms, which she accepted. We were evidently not going to return to Berkeley, so I resigned from my University of California professorship. I look back with gratitude at the forbearance of the Berkeley physics department, which had generously given me an ever-lengthening leave of absence during Louise's years at law school, in case I decided to return.

We found a wonderful house in Cambridge, which we bought from the estate of Eleanor Holmes Hinckley, a cousin of T. S. Eliot. My study in the house was in the capacious room she kept for Eliot when he visited her, as he did frequently: I am told there is substantial correspondence between them. I became ever more habituated to working at home, with my desk looking out at the garden, the television on, and often with Louise in her adjacent office, or on my office sofa.

At this time, Louise literally saved my life. Through my friendship with Bernie Feld, I found myself welcome at, and attending, international meetings of various experts on the problems of the international order. Louise understood the situation better than I did. She advised me to have nothing further to do with Bernie's world, if I wanted to get anything done in physics. She made me see that that was a world of disheartened older men giving themselves something important-looking to do, but that I was an optimistic young man with real work to do. I do not exaggerate when I confess that she saved my life.

So I even planned to end my involvement with American defense policy. But the President beat me to it. I had been acting as a consultant at the Arms Control and Disarmament Agency, studying possible Soviet anti-ballistic missile capabilities. This gave me the highest level of security clearance I have ever had, but it did not last. At some point, President Nixon decided to get rid of all outside academic consultants at the Arms Control Agency. Though this had nothing to do with me personally, for a while I was able to boast to friends that I had been fired by Nixon.

Around this time, I was asked by *Science* (the journal of the American Association for the Advancement of Science) to review a book on civil defense edited by one of my professors at Princeton, Eugene Wigner. I think of Wigner as one of the great physicists of the twentieth century. For one thing, Wigner was the first one to cut through much earlier nonsense and say clearly what in relativity theory is meant by a massive or massless particle with spin. But I did not like the book I was supposed to review. I was concerned, perhaps mistakenly, about the effect of civil defense in seeming to weaken the deterrent capability of an adversary and therefore heating up the arms race, thus injuring ourselves. I acknowledged that reasonable people could disagree about this, and now see I was perhaps too confident in my point of view. But I criticized Wigner's book because it did not even address the issue. Apart from that, I found the book to be a collection of odds and ends more or less relevant to civil defense, including chapters on shelter design, radiation biology, and the sieges of Budapest, on which in my review I acknowledged that I was no expert. Wigner later complained that he did not see why *Science* had had his book reviewed by someone who was not an expert on the sieges of Budapest.

At MIT, I continued teaching graduate courses on general relativity, which I had begun to do at Berkeley. Increasingly, now, I was emphasizing one aspect of general relativity: cosmology.

Chandrasekhar had brought particle physics to astronomy. But Einstein and his theories of space and time had brought cosmology into physics. One particular problem haunted me. Einstein had introduced a constant, Λ (lambda), into his field equations. This "cosmological constant" was the quantity of energy in space that kept the universe from collapsing from its gravitational weight, or falling apart from the propulsive force of the Big Bang. But Einstein could not figure out what it was, much less calculate some value for it. The problem haunted me.

Louise tells the story that every other day I would step out of the shower, and with a starry look report, "Honey! I think I have solved the cosmological constant!" The next day I would step out of the shower and muse, deflatedly, "No . . . ."

The spring of 1972 saw the publication of my first book, *Gravitation and Cosmology*. Over the years, the lecture notes for my courses on general relativity at Berkeley and MIT had become more substantial, gradually turning into a book. I had finally sent it in for publication in 1971. Louise says it had taken me six summers to complete. This first book of mine was competing with a new book by a formidable team of general relativists: Charles Misner, Kip Thorne, and John Wheeler. The books turned out to be complementary. Theirs goes further into the geometry of the theory, while mine has much more in it about cosmology and experimental tests of general relativity. I tried in my book to make clear that we now had a good working theory of the structure and evolution of the universe from the first few minutes to the present.

In teaching cosmology and writing the book, I was inspired by the 1965 discovery by Arno Penzias and Robert Wilson, working at Bell Labs, of a background of microwave radiation that apparently was left over from a hot early universe. Ever since the discovery in the 1920s that the cloud of galaxies filling the universe was expanding – that is, moving away from our own galaxy in all directions, it had been reasonable to suppose that there was an early phase of the history of this expansion, when matter was so condensed that there could not then have existed galaxies or stars or even atoms, only freely moving elementary particles. But it was Penzias' and Wilson's discovery of the cosmic microwave background that forced physicists to take seriously the idea that there was an early universe with conditions very different from those that we observe now.

Well, at least that was the consensus. But there were those who, like Fred Hoyle, were happier with their idea of a steady-state universe, in which matter is continually created to fill up the gaps left by the galaxies rushing apart, and on average the universe has always been pretty much the same. To the scientific mind, the philosophical attraction of this unchanging universe is substantial, because it avoids any need to suppose a "beginning" (certainly not one that religion could explain) or suggest an end (one which religion was all too ready to prophesy). But of course the question had to be settled by astronomical observation.

I recall a talk given by Hoyle in Boston sometime in the late 1960s in which he claimed that the steady-state theory was still plausible. I raised a question about the counts of radio sources with different luminosities. Those counts could be understood if on average the properties of sources were evolving with time – but not in a steady-state universe. Hoyle expressed doubts about the reliability of radio source counts. He was unwilling to give up what had seemed like a good idea.

With this fresh interest in cosmology, I began publishing papers on the subject. One was with my MIT colleague Kerson Huang, on the possibility of a maximum temperature in the early universe. (Kerson was fun to be with. I recall that I once asked Kerson why it was that, whenever we dined out together, he would always choose a Chinese restaurant. He responded with a shrug of the shoulders, remarking, "Why kid around?")

Another of my papers at that time was on the direct determination of the geometry of the universe from observations of galactic distances and spectroscopic redshifts. Another was on a kind of stickiness called bulk viscosity in uniformly expanding matter, which I thought might affect the formation of galaxies. None of this was important, but I was learning a good deal of astrophysics. Like many other physicists whose interests center on elementary particles, I would keep an eye on progress in cosmology, becoming what Murray Gell-Mann in his kindly way termed "a half-ass-trophysicist."

Later that year, I was invited to take a brief visiting chair for savants étranger at the Collège de France. MIT gave me permission to take a month off in the spring of 1971. So I accepted, and the three of us went off to see a little of France.

I gave four or five lectures on cosmology at the Collège. The audience was small, mostly confined to physicists. But I received several invitations to speak to other physics groups in the Paris area, including the laboratories at Saclay and Orsay and the Institut de Hautes Études Scientifique in Bures-sur-Yvette. At each place I spoke, the audience was pretty much limited to physicists at that place. Apparently the nature of my audiences had nothing to do with me or with the topics of my talks – it was just that French physicists then did not normally go to talks at any but their own institutions. It was very different from what I had known in the Greater Boston area, where Harvard and MIT held a joint theoretical physics seminar once a week, frequently attended also by physicists from Boston University, Brandeis, Northeastern, and Tufts. In this respect, if in no other, Boston was better than Paris.

In the autumn of 1971, I received two momentous preprints from a Utrecht graduate student, Gerard 't Hooft. In the first paper, "Renormalizable Lagrangian for Massive Yang–Mills Fields," 't Hooft argued that theories in which particles like the W and Z in the electroweak theory get their mass from the spontaneous breakdown of a symmetry are, as had not previously been understood, renormalizable. That is, in such theories, pesky infinities could be canceled. He applied this idea briefly to my 1967 model of leptons. In the second paper, "Prediction for Neutrino–Electron Scattering in Weinberg's Model of the Weak Interaction," 't Hooft continued the application of his methods to the electroweak theory, and suggested a possible experimental test of the theory. This powerful impetus was 't Hooft's essential contribution in Ben Lee's work.

It is said today that overnight, with the publication of "A Model of Leptons," the entire world of theoretical particle physics dropped everything else, and began to work on the joint enterprise of developing what would become today's "Standard Model," even before it was

FIGURE 11.1 Gerardus 't Hooft

shown to be renormalizable. There is some truth in this. My work had predicted that the W and Z particles would show up, and a boson I called the Higgs boson, as Higgs had predicted it in an earlier effort, and indeed, all of these predictions would come true. I predicted the neutral currents.

Making the profession believe me would take experimental verification. I testified before Congress on the need for the Super Collider, proposed to be built in Texas. A group of physicists at Brookhaven National Laboratory put together a paperback collection of essays explaining the need for access to an accelerator at higher energy, to which I contributed an essay, "Why Build Accelerators?" The task was close to my heart, and would become closer to my heart as I made more progress in physics. In the end, I would come to the reluctant conclusion that we had gone as far as we could go in the absence of verifying experiments at higher energies. In this early article, I took on the task of arguing that particle physics and also cosmology had a special importance among the various branches of science, as the more basic components of a grand reductionist vision. I wrote:

> Instead of feuding with one another for public favor, it would be fitting for scientists to think of themselves as members of an

expedition sent to an unfamiliar but civilized commonwealth whose laws and customs are dimly understood. However exciting and profitable it may be to establish themselves in the rich coastal cities of biochemistry and solid state physics, it would be tragic to cut off support to the parties already working their way up river, past the portages of particle physics and cosmology, to the mysterious inland capital where the laws are made.

With the failure of the Super Collider in Congress, European governments stepped in, and proposed a large collider for Conseil Européen pour la Recherche Nucléaire (CERN). I was skeptical of the progress that could be made at the Large Hadron Collider. It was simply not as large an accelerator as the Superconducting Super Collider would have been.

There remained flaws in the theory, aspects of nature it did not explain, and terms put in by hand to make it work.

So I was arguing, in all my work, both for the general public and the profession, my reductionist view, that we were engaged in a great enterprise, at the end of which we would find the ultimate laws of nature.

Louise gave me a stanza of John Donne that captured this faith:

These three hours that we have spent,
Walking here, two shadows went
Along with us, which we ourselves produced.
But now the sun is overhead;
We do those shadows tread,
And to brave clearness all things are reduced.

Of course, that my theory of leptons was renormalizable seemed like very good news. But at first, I did not trust it. 't Hooft's methods relied on the Feynman approach to field theory (known as the path integral method) that I had learned about from John Wheeler when I was a graduate student at Princeton. As I described in Chapter 5, this approach seemed to me to rely too much on Feynman's intuition, so at Princeton I had not become confident about my understanding of quantum field theory until I read Dyson's papers, which I found more sober than Feynman's.

Then, on a visit to the University of Pennsylvania, I learned that the Penn theorist Ben Lee was taking 't Hooft's work very seriously, and beginning to work on these theories. I had great respect for Lee, with whom I later became a good friend and collaborator, so I finally decided I had to look into this. I learned (possibly from Ludwig Faddeev) that in fact the Feynman approach could be justified by deduction from the accepted rules of quantum mechanics, without Feynman's hand-waving. This rendered 't Hooft's arguments much more credible for me.

A little later, the proof of renormalizability was completed in independent papers by 't Hooft and his research advisor, Martinus Veltman, and by Ben Lee and Jean Zinn-Justin. It was time to drop everything else and get back to work on the electroweak theory.

I wrote a brief paper exploring the implications of the electroweak theory for processes at high energy, where the renormalizability of the theory should prevent a runaway growth of transition rates with increasing energy. The first process I studied was the conversion of a neutrino and an antineutrino into an electron and an antielectron. No one would ever be able to measure this rate experimentally; the reason I studied it was because it had recently been examined by a formidable team of theorists: Murray Gell-Mann, Murph Goldberger, Norman Kroll, and Francis Low. Using the existing theory in which weak interactions are carried by a heavy charged W particle, they had found that, in the simplest approximation, the transition rate would grow with increasing neutrino energy at an impossible rate. They proposed to deal with this by supposing that, beyond the simplest approximations, complicated effects would damp this rapid increase. I found that, when the neutral Z particle of the electroweak theory was included in these calculations, there are lovely cancelations in the simplest approximation that keep the transition rate at high energy low, within the bounds allowed by quantum mechanics.

I also considered the conversion of an electron–antielectron pair into a positively charged and a negatively charged W particle. This can occur through purely electromagnetic interactions, which gives a transition rate that grows rapidly with energy, too rapidly to remain consistent with quantum mechanics. With the Z particle included, it can also occur through weak interactions, which also give a rapid growth with energy. But when both weak and electromagnetic interactions are included, there is another lovely cancelation, and the transition rate behaves nicely at high energy. Electroweak unification was doing its job.

In this paper, for the first time, I considered the extension of the electroweak theory to strongly interacting theories. I did not attempt a detailed theory, but only pointed out that, to preserve the renormalizability of the theory, the violations of the symmetry underlying current algebra that, among other things, gave pions their small masses would have to arise from the same spontaneous symmetry breaking in the electroweak theory that gives mass to the W and Z particles. And so it proved in the Standard Model. (In the end, we would succeed in unifying all of the forces of nature, save gravity.)

In a second paper that fall, I took a look at the experimental limits on processes that in the electroweak theory would show signs of the existence of the Z particle. Assuming the simplest possible extension of the theory to strongly interacting particles, I found that the exchange of a Z particle between a neutrino and a proton would give a neutrino–proton scattering rate in the range of 15 percent to 25 percent of the rate at the same energy of the well-known process in which a neutrino and a neutron turn into a proton and a muon. I could only give a range of possible values because the rate depended on an unknown quantity, a certain weak mixing angle that characterizes the way that photons and Z particles get mixed up with each other in the theory.

A year earlier, a group at CERN had reported an experimental "upper limit" on this ratio of between 6 percent and 18 percent. This was odd. Usually experimenters give a probable range of their results only when they are actually seeing something; upper limits on something that is not seen are usually reported as a single number. I guessed that they were actually seeing neutrino–proton scattering events, but having not expected to see them they reported their rate measurements as an upper limit. It would take a few years to settle the matter.

In one other paper, I joined forces with my friend Roman Jackiw, a younger colleague at MIT. Roman had an ebullient personality, which I described in an anecdote contributed in 2020 to a Jackiw festschrift:

> Roman and I were at some conference or other in San Francisco. Somehow we and our wives found ourselves together downtown. It was getting on toward midnight, and we were about to part reluctantly, when Roman said that we couldn't say goodnight because we had to go to the Ritz Old Poodle Dog. What is that? Why, a restaurant of course. But why do we have to go? Well, of course, because it was a great San Francisco eatery, and the King Crab was just in. What was that about the King Crab? Why it only comes in to San Francisco in June. But Roman, it is nearly midnight. Of course, that's when you have to go to have King Crab at the Ritz Old Poodle Dog. Convinced by this logic, we arrived at a magnificent establishment, and the waiter said in reverent tones, "You have come for the King Crab." There we were at midnight, the four of us, having the time of our lives. Not to mention a couple of hours of great conversation, and some fabulous crabs on ice. With champagne as Roman said "of course."

What a delightful companion he was!

There had been very accurate measurements of the strength of the magnetic field produced by muons, sensitive enough to detect the effect of emission and absorption of photons by the muon. Experiment and theory were in good agreement. Jackiw and I checked whether in the electroweak theory the emission and absorption of Higgs bosons would disturb this agreement. We found that, because of the weakness of the interaction between Higgs bosons and muons, the effect of

Higgs bosons would be too small to be detected. The result was a little disappointing, but it was exhilarating that now, with a renormalizable theory, we could do the sort of calculations for weak interactions that were previously only possible in quantum electrodynamics.

The electroweak theory was beginning to receive a good deal of attention from theorists. Every February, a small meeting of physicists was held at Kitty Oppenheimer's house in Princeton to commemorate the anniversary of her husband Robert Oppenheimer's death in February 1967. At this meeting in February 1972, I was asked to give one of the talks on current research, to cover the electroweak theory. After my talk, the discussion leader (perhaps Robert Serber) called on the next speaker, but Freeman Dyson objected, saying that he wanted to hear more about the electroweak theory. It was the moment in my life that I felt most clearly that I had made a mark in physics.

I had planned that, after the meeting, I would drive Louise to Newark Airport for our flight back to Boston. A heavy snowfall made that impossible. Instead, for the only time in our lives, we took a train all the way from Princeton Junction to Boston. We sat with Ludwig Faddeev, who had been at the meeting in Princeton, and had many interesting remarks about field theory and Russia. Viki Weisskopf was also with us. We four were seated in an oddly old-fashioned compartment, like the ones in Hitchcock's thriller *The Lady Vanishes.* Riding for many hours with a voluble Russian through a snowy landscape, in such a compartment, we felt transported into a novel by Tolstoy or Dostoyevsky.

The birth of the term the "Standard Model" came about in the following way. I did not want to elevate this theory to a dogma, so I began to call it the "Standard Model," and tried to explain that, of course, the Standard Model might be partly or wholly wrong. However, its importance lay not in its certain truth, but in the common meeting ground that it provides for an enormous variety of astronomical data. By discussing these data in the context of a standard cosmological model, we can begin to appreciate their cosmological relevance, whatever model proves correct. It was only a few years later that the term the "Standard Model" began to be used by me and others to describe our modern theory of elementary particles and their interactions, in much the same spirit in which I had used it in cosmology.

I didn't realize it at first, but my book on cosmology would be competing with a book on cosmology by one of the century's leading cosmologists, Jim Peebles. In his Nobel lecture, Peebles said that his book was more complete in the consideration of astrophysical processes, while mine was more complete in the considerations of mathematics. I think that was fair. As he said, "the two books signaled the change in physical cosmology from its near dormant state in the early 1960s to the start of a productive branch of research in physical science in the late 1960s."

The publication of this book began a change in my life that would eventually end my active involvement in JASON. I found I liked writing books. Technical books like this would not make me rich, but the royalties could replace consulting fees as a supplement to our regular salaries. Importantly, since no classified material was involved, I could write books at home. I loved listening to music or watching television while working. I loved working in my study (formerly T. S. Eliot's bedroom). And I admit that I was not immune to the pleasure offered by good book reviews. Once in 1972, I was browsing at a magazine rack in Reading International, a book shop (alas, now gone) at the corner of Brattle and Church Streets in Cambridge. In the British popular science journal *New Scientist*, I came across a review of my book, in which I read "What a book!" No one had ever said "What a report!" about any work I had done for JASON.

That spring, I learned that I had been elected to the National Academy of Sciences. I suspect that I had been nominated by Viki Weisskopf. The National Academy's activities are centered in Washington, so it has had less importance in my life than the American Academy. Still, I was naturally very pleased to have my work recognized in this way while I was still in my thirties.

At this time, I was offered a professorship at Harvard. Julian Schwinger was leaving for UCLA, where he could play tennis all year, leaving a very distinguished open slot. I was attracted by the idea of moving to Harvard. For one thing, from our house, I could walk to work across the Cambridge Common, while to get to my office at MIT, I had to drive to a parking garage and then walk through a long corridor in the nutrition department that always smelled of soup. Also, in much the same way that I liked looking at the old stone walls in Belmont and living near the Cambridge Common where Washington had marshaled his troops, I liked the idea of being at America's oldest university. But I did not see how I could accept. People at MIT had been very kind to me, and since the physics department at MIT was every bit as strong as the one at Harvard, I had no creditable reason for the move.

Then one evening, I received a phone call from Paul Martin, who was at the time chairman of the Harvard physics department. They were upping the ante. Not only would I fill the empty slot left by Schwinger's departure, I would have his endowed chair, the Higgins Professorship. This incidentally came with Schwinger's wood-paneled office, with its marble fireplace, perhaps the only fireplace in any Harvard professor's office. (Eugene Higgins was an eccentric millionaire with no connection with Harvard, who on his death was found to have left his fortune for the endowment of chairs in physics at Harvard, Princeton, and I think also Columbia and Yale, though not at MIT. Murph Goldberger was the Higgins Professor at Princeton.) I would also be a senior scientist at the Harvard-Smithsonian Center for Astrophysics, which was clustered along with the Harvard College Observatory and the Department of Astronomy on Observatory Hill, not far from our house. I thought that I could depart from MIT in good conscience. Anyone could see that this offer was too good to turn down. I recall that, after Paul Martin's call, I looked out of the back window of our house at beautiful autumn leaves, and felt a little chill that now I knew where I would spend the rest of my life. On this last point, I turned out to be wrong.

12 Cambridge: 1972–79

Many Americans who lived through the 1970s remember it as a bad decade, a time of "stagflation" – high inflation and high unemployment. Then there was Watergate. There was the oil embargo. But for physicists who study elementary particles and quantum field theory, it was a golden age. For us, those glory days saw the experimental confirmation of the electroweak theory, and the extension of that thinking would lead us to a successful theory of strong interactions as well. All the fundamental forces of nature, except for gravity, would be unified in what became known as the "Standard Model."

In the winter of 1972, before I made the move from MIT to Harvard, I took the family with me on a trip to Coral Gables for a conference on unified theories in physics and astrophysics, hosted by the University of Florida. I was to give a talk summarizing the recent progress in the theory of weak and electromagnetic interactions. With Ben Lee and his family, we took the occasion to visit the Parrot Jungle, a nostalgic experience for me, remembering seeing it with my parents more than three decades earlier.

During the Coral Gables meeting, several physicists mentioned to me that it was a pity that I was not going to receive the Oppenheimer Prize. This was a surprise to me, because not only did I not know that I was being considered for the Oppenheimer Prize – I had not known that there was an Oppenheimer Prize. Looking into this interesting matter, I found that, indeed, there was an Oppenheimer Prize granted in these annual conferences at Coral Gables, and that I had been named by a nominating committee to receive it that year. The nomination was then vetoed by Eugene Wigner and Edward Teller, as I was told, apparently because of their disapproval of my positions on missile defense and civil defense. This all came out in public when Murph Goldberger (ever my guardian angel) stood on the dais and told the meeting that it was wrong that this prize was denied to me on political grounds. Sitting in the audience, I felt happier with Goldberger's remarks than I would have felt on receiving the prize. I did receive it, a year later. I was told that Wigner had been brought round (perhaps by John Wheeler). It was my first award.

I moved to Harvard in the autumn of 1973. I was assigned the very nice office with a fireplace. On the floor of the closet in my new office, I found a pair of shoes left there by Julian Schwinger, as if daring me to fill them. I asked Sidney Coleman what he thought I should do with them. Without missing a beat, Sidney replied, "Bronze them."

Especially at that time, Harvard was a good place to be in physics. With a generous research contract, the National Science Foundation was supporting at Harvard a group of theoretical particle physicists in which the senior faculty (besides me) were Sidney Coleman and Shelly Glashow, backed up by an exceptionally talented cadre of junior faculty, visitors, postdocs, and graduate students, some of whom the reader will meet as we go along.

After it had been proved that renormalization could tame the infinities in theories in which photon-like particles get mass from spontaneous symmetry breaking, it was widely believed that some such theory could correctly describe the weak interactions. Whether the specific electroweak theory proposed by me and Salam was the correct theory was an open question, for me as well as for other theorists. We had simply used it as an illustrative example. At the time, I would not have much minded if the original electroweak theory had played that role. It would have pointed the way, at least, toward a correct theory.

Indeed, the electroweak theory had a regrettable feature, in that it involved an unknown quantity, the weak mixing angle mentioned in Chapter 11. So the theory could predict only a range of possible results for any experiment. Late in 1971, I had described a theory of weak and electromagnetic interactions in which the electroweak theory appeared as an approximation to a larger theory, but with a fixed value of 30° for that angle. With a Harvard Junior Fellow, Howard Georgi, Shelly Glashow even proposed a theory with no Z particles and no undetermined angle, which I thought was an interesting possibility. (Howard Georgi later became a tenured faculty member at Harvard and a leading figure in particle physics.)

Covering all my bets, in the autumn of 1972, I had set out the general formalism for any theory of weak and electromagnetic interactions in which the photon-like particles that transmit the weak interactions get mass from spontaneous symmetry breaking. It was up to experiments to decide which was the correct theory. Clearly the issue could be decided by the discovery and measurement of the new "neutral current" weak interactions, carried by the electrically neutral Z particle of the electroweak theory. It also seemed obvious that the best way to do this would be to search for the scattering of a high-energy neutrino exchanging a Z particle with a proton or neutron in an atomic nucleus. Beams of high-energy neutrinos produced in the decay of pions into muons and neutrinos had become available, and had been used in detailed studies of the process in which the collision of a neutrino with a neutron or proton yields a muon and a proton or neutron, presumably due to the exchange of a W particle.

As I mentioned in Chapter 11, experiments at Conseil Européen pour la Recherche Nucléaire (CERN) may have been detecting the effects of Z exchange neutral currents in neutrino scattering by nuclei. But, not expecting to see such events, physicists had reported only an upper bound on the scattering rate.

At that time, the faculty member at Harvard who was most active on the experimental side in high-energy particle physics was Carlo Rubbia, with whom I had briefly overlapped at Columbia. Carlo typified a new kind of physicist that has been brought to prominence by the huge increase in the size of research groups in experimental elementary particle physics. Some of their papers list hundreds of physicist authors. In addition to the technological skill and understanding of theory that experimenters have always needed, Carlo had an almost military commanding presence.

I urged Carlo to set in train a search for the scattering of neutrinos by nuclei, which would provide evidence for the existence of the Z particle. In 1973, the Harvard–Pennsylvania–Wisconsin–Fermilab group, Rubbia among them, found such scattering events, but publication of their paper was delayed. They took the opportunity to rebuild their detector, and at first did not find the same signal.

As it happened, the first clear evidence of neutral currents was provided by a different group. At the end of June 1973, a group at CERN reported a single event in which a neutrino from the decay of a high-energy pion into a muon and a neutrino was scattered by an electron. This caused a recoil of the electron that produced a visible trail of bubbles in a bubble chamber.

This was something that could occur through the exchange of a neutral Z particle, but could not happen in a theory of weak interactions with just charged W particles. With a single event, it was not possible to draw precise quantitative conclusions. The CERN experimenters were able to calculate only that the rate of the scattering process was consistent with the electroweak theory with a weak mixing angle between 20° and 50°.

The rate for neutrino–electron scattering is small because it is proportional to the square of the energy available in the collision. And since electrons are so light, most of the initial neutrino energy is not available for the Z exchange process, just going into the recoil of the electron. Protons and neutrons are almost two thousand times as heavy as electrons, so much less energy goes into the recoil of the particle hit by the neutrino. More energy is available in the interaction, and the scattering rate is correspondingly much higher.

Unfortunately for the Harvard–Pennsylvania–Wisconsin–Fermilab group looking for this reaction, in the summer of 1973, they were scooped by the group at CERN. But the neutrino beam at Fermilab had higher energy than the beam at CERN, so a little later that summer the Fermilab group were able to give a moderately precise result, that the ratio of scattering of their mixture of neutrinos and antineutrinos on atomic nuclei, to the W exchange process in which the neutrino or antineutrino turns into a negative or positive muon, is between 20 percent and 38 percent. Depending on the unknown weak mixing angle, the 1967 model of leptons would give a ratio between 22 percent and 55 percent. The experimental results were consistent with the theory, and showed clearly that the "neutral current" events of the sort produced by Z exchange do exist. But it was hardly a ringing endorsement of any specific theory. By the end of 1973, the CERN group was able to give separate results for the scattering of neutrinos or antineutrinos by nucleons, and verified that both scattering rates were consistent with the same value of the weak mixing angle, between 33° and 39°. Being able to fit two observed rates with a single choice of angle was at least a tentative verification of the electroweak theory.

Despite the impartiality that is expected from scientists, I began to feel that I had some stake in the validity of the electroweak theory. In a television interview, Richard Feynman was asked what he thought about this theory. I was disappointed to hear him say that he did not believe it. What bothered him was just then what was bothering me and other theorists. In essence, the awkward thing was that the theory had an undetermined quantity, the weak mixing angle. I wrote to Feynman acknowledging this, but explaining that this is just the sort of theory that one might expect to find as the limit at relatively low energies of some more attractive theory. I received no reply.

Some months later, late at night, we were awakened by the shrill ring of my telephone. My home office in T. S. Eliot's big old bedroom was adjacent to our own little bedroom, and I bumped my way in the dark to my desk. The call was from Feynman. (He may have forgotten the three-hour time difference between Pasadena and Cambridge.)

When I bumped my way back, Louise was awake, and sleepily asked who that was. I said, "It was Dick." She said, "Good grief, what did he want?" What Feynman had said was, "At last I believe you."

Feynman had even given me a brief reason for this change of heart, but I didn't understand the explanation. Feynman had his own way of looking at things, and would not believe anything unless he could fit it into his own worldview. But of course I was glad that he had managed to fit in mine.

About a month before the first detection of a neutral current/weak interaction event, there had been an exciting theoretical breakthrough regarding the strong interactions. Experiments on reactions produced in the collisions of high-energy electrons with neutrons or protons had earlier shown that strong interactions seem to become weaker at high energy. (This was known as "Bjorken scaling.") It was already understood that effective interaction strengths can depend on energy. For instance, back in 1954, Gell-Mann and Low had calculated how the effective electric charge with which particles such as electrons attract or repel each other varies with the energy at which they collide. But they had found that, with increasing energy, the effective electric charge increases rather than decreases. The same was found to be true of interaction strengths in most theories. Then, at the end of April 1973, several theorists identified a class of theories in which the interaction strength decreases with increasing energy, as indicated by experiments. These are theories that, like the electroweak theory, are of the Yang–Mills type, in which photon-like particles carry the charges with which the same particles interact. The theorists who found that the charges decrease with increasing energy in these theories were David Gross and his PhD student Frank Wilczek at Princeton, and Coleman's PhD student Hugh David Politzer at Harvard. This phenomenon is known as "asymptotic freedom," because these theories of strong interactions at high energy asymptotically become theories of free particles.

These results pointed naturally to a specific theory of strong interactions. In 1964, Murray Gell-Mann and George Zweig had independently proposed that all strongly interacting particles, such as protons, neutrons, and pions, are composed of more truly elementary particles, which Gell-Mann called "quarks," and their antiparticles.

The various types of quark are distinguished by an attribute known as flavor. Protons and neutrons are composed of quarks of two different flavors: There are up-quarks of charge $2e/3$ and down-quarks of charge $2e/3$ (where "e" is the charge of the electron). Protons consist of two up-quarks and a down-quark; neutrons consist of two down-quarks and an up-quark. Pions consist of a quark and an antiquark. Quarks with other flavors are present in some heavier unstable strongly interacting particles.

Several theorists had further proposed that the quarks of each flavor come in three varieties, fancifully called "colors." It was natural to identify these colors as the couplings, analogous to electric charges, with which the photon-like particles interact in the asymptotically free theories studied by Gross and Wilczek, and Politzer. By analogy with quantum electrodynamics, but with color taking the place of electric charge, this theory became known as "quantum chromodynamics."

This was all very interesting, but if these photon-like particles have zero mass and interact strongly with the quarks inside protons and neutrons, why had they not been seen experimentally years earlier? The growing acceptance at that time of the theory of weak and electromagnetic interactions, in which photon-like particles get masses from a spontaneous symmetry breakdown, led at first to the supposition that much the same occurs in quantum chromodynamics. In other words, the photon-like particles in this theory acquired masses from a spontaneous breakdown of the symmetry among the three colors, and had not been produced experimentally because they are too heavy.

Then I had a different idea. About a month after Gross and Wilczek, and Politzer announced their exciting results, I proposed that the color symmetry underlying quantum chromodynamics is not spontaneously broken. Instead, the photon-like particles in this theory are actually massless, like photons. But their interactions become so strong when they are pulled apart from other colored particles that they can never be seen in isolation. The same suggestion was made independently a little later by Gross and Wilczek.

This suggestion of color trapping had the great merit of also offering an explanation of why quarks had not been observed. Quarks were expected to have mass, and it might have been supposed that they are so much heavier than nucleons (that is, protons and neutrons) that they could not yet have been produced in high-energy collisions. But if nucleons are composites of these very heavy quarks, then the quarks inside a nucleon would have to feel very strong attractive forces to cancel their masses. Yet the successes of the quark model depended on their not feeling strong forces inside a nucleon. With the new perspective of color trapping, it was supposed that quarks are not particularly heavy, and, like all particles carrying color, feel very strong forces only when one tries to pull them away from other colored particles.

This view of quantum chromodynamics had another advantage that I found especially attractive. A renormalizable theory involving only quarks and photon-like particles that interact with the color carried by quarks and themselves has to be very simple. It has to be so simple, in fact, that it has no way to distinguish right from left, particles from antiparticles, or time flowing forward or backward. All these symmetries had been known to be respected by the strong and electromagnetic interactions, but were violated by the weak interactions. So they could hardly be fundamental principles of nature. Now they were revealed to be accidental symmetries – a consequence of the great simplicity of quantum chromodynamics and quantum electrodynamics, imposed by the condition of renormalizability.

This condition of renormalizability is respected by a shift in the values of various physical constants. (The same was true of the conservation of the various quark flavors.) But as I now emphasized, this condition could not be satisfied if the theory involved additional strongly interacting spinless fields whose values in nature would break the color symmetry, and thereby give masses to the photon-like particles of quantum mechanics – in the way that W and Z particles get masses in the electroweak theory.

There was still little in the way of experimental confirmation of quantum chromodynamics or the electroweak theory, but by the start of summer 1973, it seemed to me, as a theorist, that things were clicking nicely into place.

I had an invitation to give a plenary talk at an international conference on high-energy physics at Aix-en-Provence at the end of the summer, where I intended to outline what we had learned in that magical spring.

First, however, I was taking my family to La Jolla. This was to be my last participation in a JASON summer study, although I continued for years to go to its spring and autumn meetings in Washington.

On the way back to Cambridge from La Jolla, we stopped in Colorado, where I had been invited to join a roundtable conference at the Aspen Institute. The multidisciplinary group included literary critics Lionel and Diana Trilling, historian Lord Alan Bullock, author Saul Bellow, and Mortimer Adler – all of whom were engaging, although Bellow was rather silent, and Adler rather bumptious. The only scientists at the table were the chemist Sir Frederick Dainton, chair of the British University Grants Committee, and me. I wish that I could remember saying anything brilliant. The one remark of anyone's that I do remember was Bellow's quote of E. M. Forster, who had said that he writes to earn the respect of those he respects, and to earn his bread. Not bad motives for anyone.

I planned to take the family to Oxford for a month, before giving my talk at Aix. We would stay at the old Randolph Hotel, which we had liked in a brief visit in 1962. I could prepare my talk in the cool English summer, and we could see a little more of the English countryside.

After our month in Oxford, we flew to Marseilles, and drove (thankfully, on the right) to Aix-en-Provence.

In my talk, "Recent Progress in Gauge Theories of the Weak, Electromagnetic, and Strong Interactions," I projected the outlines of what later became known as the Standard Model of elementary particle physics. I expressed confidence in the general idea of a gauge theory of weak and electromagnetic interactions based on a spontaneously broken symmetry, and pointed to the encouragement provided by the discovery that year of neutral current neutrino interactions. But I was cautious about identifying the electroweak theory of myself and Salam as the realization of these ideas, particularly chosen by nature. I also expressed confidence in the gauge theory of strong interactions, citing "the exciting new thing that has happened recently," the discovery of asymptotic freedom by Gross and Wilczek, and Politzer. I emphasized how in this theory one could understand the mystery of symmetries being respected by strong and electromagnetic interactions but not weak interactions.

This was for me the most exciting talk I have ever given, because I had the feeling that many physicists in the audience were appreciating for the first time the enormous progress that particle physics had made in the past few months. In the evening, after my talk, I sat with Louise and Elizabeth outside in a café on the Cours Mirabeau, chatting with conference participants who stopped by our table, and feeling wonderful.

That autumn, Harvard was opening its new Undergraduate Science Center. I was invited to give a talk at the November inauguration of the center. The invitation came from my old friend from Berkeley days, George Field, a leading astrophysicist now at Harvard. I gave a "popular" cosmology talk, describing what we now know about the early universe. I talked about the falling temperature and density of the particles filling the universe from an early time, when there were only freely moving protons, neutrons, electrons, and photons, until the binding of the first atomic nuclei, of deuterium and helium, at the end of the first three minutes. I made a wisecrack that, after the first three minutes, nothing of interest would ever happen in the history of the universe. Of course, that was not true even from the limited perspective of an astrophysicist, but it got a laugh.

At some time after this talk, I received a call from Erwin Glikes, who introduced himself as the editor in chief of Basic Books. He had heard of my talk from a mutual friend, the sociologist Daniel Bell, who had been in the audience at the Undergraduate Science Center. Glikes urged me to expand this talk into a book on cosmology for a general audience. He thought that with my subject and especially the title, *The First Three Minutes*, the book would sell well. I was very busy with elementary particle research, and so at first I declined. But the suggestion did not leave my mind.

In part because of my new connection with the Harvard-Smithsonian Center for Astrophysics, I took some time now to work on astrophysics. It was known that, in the collapse of a star in a supernova explosion, enormous floods of neutrinos are emitted, as protons absorb electrons, thereby turning into neutrons. (Neutrinos from a supernova explosion were first detected in 1987.) As I mentioned in my talk at Aix-en-Provence, in the electroweak theory, these neutrinos can be scattered by matter in the outer layers of the star, giving up some of their momentum to the matter. The big problem was to find under what conditions this scattering would eject the outer layers of the star, producing a supernova rather than a black hole.

I read Chandrasekhar's book *Radiative Transfer* to learn analytic techniques that I could apply to neutrinos instead of radiation. Unfortunately, my analytic efforts turned out to be a sampan confronting a battleship. James Wilson, using the enormous computing capability at the Livermore National Laboratory (where they design nuclear weapons), was able to do detailed calculations of the effects of neutrinos in supernovas. His calculations blew my feeble analytic efforts out of the water.

I turned to a problem more suitable for analytic treatment. I had seen a preprint, by D. Kirzhnits and Andrei Linde, both working at the Lebedev Physical Institute in Moscow, on the restoration of spontaneously broken symmetries at high temperature. It was well known that both the spontaneously broken symmetry between different directions in a ferromagnet and the spontaneously broken "gauge" symmetry in a superconductor are restored at a sufficiently high temperature. Kirzhnits and Linde showed how the same thing could happen in a relativistic quantum field theory. This could be relevant in the very early universe, when the temperature was high enough to affect the fields of the Standard Model. Reading the Kirzhnits–Linde preprint, I realized that I had a lot to learn about quantum field theory at high temperatures. After some self-education (which was to serve me well a few years later), I was able to show that this sort of symmetry restoration could occur in the kind of theory of photon-like particles used in the Standard Model.

In those days, the seminars of the Harvard theory group were held in what was called the faculty room, a wood-paneled room in Lyman Laboratory, just large enough to hold us all. One day, during a seminar talk, I heard an unusual slurping sound from somewhere in the audience. Looking around, I saw that one of our postdocs was nursing her baby. I couldn't help smiling. After spending years in exclusively male scientific environments, this was vivid evidence that, at least here and there, opportunities were opening for women, and people were becoming more accepting. For years after I joined JASON, for example, it had no women members. Women were wives and secretaries. After Louise became a lawyer, one of the JASON wives told her, "You were the only one of us to escape." Today, the chairperson of the JASON group is a woman.

Louise recalled that, one day, during one of our return trips to Stockholm, in a brief gathering of the wives, one woman had said, pointing to the suit Louise was wearing, "Poor thing! She has had to get by in a man's world." Louise remembers with some satisfaction that she had smiled and said, "Haven't you?"

One day Louise and I visited the observatory on Palomar Mountain. Our host, Maarten Schmidt, telling Louise to wait standing in an unshaded path, started to guide me into the observatory. I had the sense to beckon Louise to accompany us. Schmidt seemed annoyed. In the observatory, Schmidt allowed me to see what it was like sitting in the observer's cage. He apologized to Louise, telling her that it was a rule at Palomar, laid down by former director Horace Babcock, that women could not enter at all, he had let her in unthinkingly, and she could not be allowed in the observer's cage. I do not know why I put up with this, and she does not know why she put up with it, but we were creatures of our times, and times were very different then. Today, I would have refused to enter without my wife, and would have turned on my heels and left if she were treated in such a way. This openly debasing treatment of women was particularly disturbing in astronomy. Throughout the twentieth century, women had been making important contributions to astronomy – one thinks, for example, of Henrietta Leavitt, who made it possible to measure the distances of galaxies participating in the expansion of the universe. Today, Babcock's rules are history, and women like Sandra Faber are among the world's leading astronomical observers.

At some time early in 1974, I was in the faculty room listening to a talk on quantum chromodynamics, probably by Politzer. On the blackboard was the curve showing the very gradual decrease, with increasing energy, of the quantity known as a coupling. This decrease characterizes the strength of the strong interactions. In my mind's eye, I could see the curves showing the even slower decrease with energy of one of the two couplings that characterized the strength of the electroweak interactions, and the slow increase of the other. It suddenly occurred to me that, at some very high energy, these three curves might all come together – not necessarily with the couplings equal, but with the ratios of the couplings taking specific values, not very large or very small, as dictated by some symmetry uniting the electroweak and strong interactions.

As it happened, just such a symmetry had been proposed a little earlier by Georgi and Glashow. It had at first seemed to be ruled out by experiment. The weak angle was predicted to be about 37.76 ... °, which was not favored by recent data on neutrino scattering. Worse, the strong interactions in this theory were not much stronger than the electroweak interactions. Worse yet, the theory violated conservation laws that keep ordinary matter stable, so that – for example – a proton could decay into a pion and a neutrino. A different version of a symmetry uniting strong and electroweak interactions had been offered independently by Salam and his collaborator, Jogesh Pati, and Georgi soon offered another attractive theory, but these had most of the same problems. A large class of theories uniting the strong and electroweak interactions predicted the same ratios of coupling constants as the Georgi–Glashow, Pati–Salam, and Georgi theories.

But these problems would go away if the symmetry uniting the strong and electroweak interactions was broken at a very high energy, much higher than the energies accessible in any laboratory. This would be the energy where the one strong and the two electroweak couplings would have the ratios dictated by the symmetry. The couplings at accessible energies would have to be calculated by following the increase of the strong coupling and the slower changes of the electroweak couplings as the energy decreases toward the range in which laboratory measurements are made.

To do this calculation, I joined forces in the spring of 1974 with Georgi and an assistant professor, Helen Quinn. They had expertise in symmetry applications and the energy dependence of coupling constants, and probably had already been thinking along similar lines. We found that, in order for both of the two ratios of the three couplings of the Standard Model to reach the values dictated by a large class of possible symmetries uniting the electroweak and strong interactions at some single high energy, the weak angle at laboratory energies would have to be not 37.76 ... °, but (here using more recent data on strong interactions) close to 27°, in good agreement with the experimental value. The strong interactions are stronger than the electroweak interactions at ordinary laboratory energies. As the energy at which these interactions are felt decreases from the unification scale to the laboratory scale, the strong interactions get stronger while the strengths of the electroweak interactions hardly change.

Strikingly, the energy at which the ratios of the coupling constants reached the values dictated by the symmetry was enormously high. Using current data, this calculation gives a unification energy of about 10^{15} GeV! (1 GeV is the energy given to an electron by a billion-volt electric battery. It is close to the energy contained in the mass of a hydrogen atom. 10^{15} is a one followed by fifteen zeros.) The rate of decay of ordinary matter, as for example the decay of a hydrogen nucleus into a pion and a neutrino, would be inversely proportional to the fourth power of the energy at which the symmetry between electroweak and strong interactions is broken. So with this energy, as high as 10^{15} GeV, the worry about the stability of matter disappears. Not only would the mean life of ordinary atoms be longer than the age of the universe; it would be so long that it would be difficult to see even a single atom decay in a large laboratory sample of matter.

This calculation revealed what came to be known as the hierarchy problem: What accounts for the enormous disparity between the energy at which the symmetry unifying electroweak and strong interactions is broken, and the energies (in the general neighborhood of 1 GeV, give or take a few powers of 10) contained in the masses of known particles? But there always had been a hierarchy problem. From Newton's constant of gravitation, the speed of light, and the Planck constant of quantum mechanics, it is possible to calculate a single quantity with the units of energy. It is known as the Planck energy, about 10^{18} GeV. This is roughly the energy in the mass that the electron would have to possess for the gravitational attraction between two electrons to equal their electrostatic repulsion. The fact that the unification scale of 10^{15} GeV is less than the Planck scale means, in retrospect, that Georgi, Quinn, and I had been justified in ignoring gravitational effects in our calculation. But it is intriguing that these two scales are not very different. No one knows why.

In November 1974, physicists received exciting news of an unusual new particle, discovered simultaneously by Sam Ting's group at Brookhaven National Laboratory, and Burt Richter's group at the Stanford Linear Accelerator. Ting and Richter gave different names to the particle, but it has become known as the J. The J particle weighed about three times as much as a proton, and yet was nearly stable.

There were, at the time, three known types of quarks, the up- and down-quarks that make up protons and neutrons, and a third strange quark contained in unstable particles like the K meson. None of these three quarks was heavy enough to account for the mass of the newly discovered particle. I recall that, at a meeting at MIT to discuss this discovery, Viki Weisskopf said that no one had any idea what this particle could be – and that that was what made it so exciting. For once, Viki was wrong. Theorists at Harvard had a pretty good idea what it was. That summer Politzer and a Harvard postdoc, Tom Appelquist, had explored the possibility of a particle composed of a new heavy quark and its antiparticle. Because in quantum chromodynamics the strong interaction becomes weaker at high energy and at small separations, the annihilation of the quark and antiquark would be relatively slow, leading to a nearly stable bound state. A heavy fourth quark had been proposed in 1970 by Glashow, with Luciano Maiani and John Iliopoulos, as a means of canceling what would otherwise be an excessive rate of processes like the decay of K mesons into muon–anti-muon pairs produced by the exchange of pairs of W particles. When I extended the electroweak theory to quarks in papers in 1971 and 1972, I had included the fourth quark along with the other three, and found that this also avoided the rapid decay of K mesons into muon–anti-muon pairs, produced by the exchange of a single Z particle. So to some of us it seemed clear that the new particle was a bound state of the fourth quark and its antiquark. We saw its long life as a lovely confirmation of the weakness of strong interactions at short distances embodied in the asymptotic freedom of quantum chromodynamics.

And so it proved. Not only was there a fourth quark, but before long, particles containing even heavier fifth and sixth quarks were discovered. Because of the asymptotic freedom of strong interactions, the masses of the new quarks were easily calculated from the observed masses of the new particles that contained them.

Still, no one has been able to understand why these quark masses are what they are. The six quarks form three doublets, each with one quark of charge $2e/3$ and another of charge $e/3$, like the up- and down-quarks. For each quark doublet, there is a doublet of particles called leptons, which do not feel the strong interactions. There is one lepton of charge e, like the electron, and a corresponding very light neutral neutrino. Each matched pair of quark and lepton doublets is called a generation.

We do understand why the quark and lepton doublets have to match. In 1969, John Bell and Roman Jackiw, and independently Steve Adler, had discovered subtle effects that can destroy the apparent symmetries of a theory. As I and others pointed out in 1972, the matching of quark and lepton doublets in each generation is necessary for the cancelation of effects of this sort that would destroy the symmetries on which the electroweak theory is based. But we have no idea why there are three generations, rather than one or two, or a hundred.

In the spring of 1975, I returned for a while to current algebra, now in the context of quantum chromodynamics. The symmetry on which the current algebra of the 1960s was based is automatic in the limit of zero up- and down-quark masses, and is therefore a good approximation, if these masses are relatively small. But in this case, in quantum chromodynamics, there apparently is another accidental symmetry. I showed that the spontaneous breakdown of this additional symmetry necessarily leads to a neutral particle with a mass less than the pion mass times three. But no such particle exists. For a while, this was for me the outstanding unresolved problem with quantum chromodynamics, until in 1976 Gerard 't Hooft showed that the interaction of quarks with certain twisted configurations of gluon fields violates the apparent symmetry of quantum chromodynamics that led to this unwanted light neutral particle. So quantum chromodynamics was in no trouble after all.

I had an invitation to give that year's Scott Lecture at the Cavendish Laboratory in the other Cambridge.

That summer, I found myself driving out my daughter for riding lessons. This turned out to be a good thing for me, because I had to read something while waiting for Elizabeth, and so I took the opportunity to study a new review article, published the previous year by Kenneth Wilson and John Kogut. This article laid out the theory of critical phenomena. These are phenomena in which, at a certain temperature or pressure, there suddenly appear correlations of long range, like the correlation of the directions of the spins of distant atoms in a ferromagnet. I became interested in critical phenomena because the mathematical methods used in studying it were similar to those used in particle physics, for instance in calculating the change of the strong interaction strength with energy in quantum chromodynamics. Also, I knew that Wilson and his Cornell colleague, Michael Fisher, had scored a great success in 1972 using these methods to calculate the way that magnetism depends on temperature, near the temperature at which magnetism disappears. What I learned during her lessons turned out to be important to me later in thinking about the quantum theory of gravitation.

In the following autumn, Louise received a non-lookover invitation to visit Stanford Law School in the 1976–77 academic year. Stanford is so isolated in Palo Alto that much of its faculty is away visiting somewhere else; so it conducts an extensive invitational program as well.

To keep Louise company, I arranged to visit the Stanford physics department, and we enrolled Elizabeth in Castilleja School in Palo Alto. We planned that, before heading out to California, and after completing our 1975–76 academic year, we would visit Sicily, where I was to lecture at the Erice (pronounced "Ehritchay") summer school.

Meanwhile, throughout 1976, the experimental evidence from neutrino scattering kept strengthening the case for the original electroweak theory of Salam and myself. As I admitted in Chapter 11, at first I would not have minded much if it had turned out that this theory was just an illustrative example of the class of theories of weak and electromagnetic interactions, which included whatever would be revealed by experiment as the correct theory. But I confess that, with the growing experimental support for our original electroweak theory, I was beginning to feel some personal stake in that theory. This feeling was hardly diminished when a friend sent me an issue of the Stockholm newspaper *Dagens Nyheter*, which predicted that Salam and I would receive the Nobel Prize in physics that year, for the electroweak theory.

At that time, physicists seemed to be waiting for one more specific test of the theory, having to do with the weak interactions of electrons rather than neutrinos. The interactions of electrons with each other and with nucleons is dominated by electromagnetic forces, but these forces respect space-inversion symmetry, the invariance of the equations under reversal of all directions in space, while the exchange of a Z particle between an electron and a nucleon does not. This produces a mixing of states of electrons in atoms that would otherwise be separated by their different behavior under space inversion. Physicists at Oxford and at the University of Washington in Seattle realized that this mixing in atoms of bismuth would result in a rotation of the plane of polarization of light as it passed through bismuth vapor. At the time that I took my family to Erice in the summer of 1976, rumors were circulating that this effect had been seen.

Erice is an ancient town, known as Eryx to the Romans, but it is now dominated by buildings of the eighteenth century, dozens of them churches. At an elevation of 2,500 feet, and with breezes blowing in from the Tyrrhenian Sea to the west, it is wonderfully cool, even in the hot Sicilian summer. The streets are cobbled and narrow, too narrow for even one car. It was rather like being on the set of the opera *Cavalleria Rusticana*. I gave a minicourse on critical phenomena, based on what I had read in the Wilson–Kogut article and other reading, reworked to be accessible to particle physicists like me.

The Erice summer school was run by the experimental physicist Antonino Zichichi. He had a reputation of knowing everyone in Italy. A joke went around the summer school, according to which a visitor to Italy was told that Zichichi was friends with the pope. To test this, the visitor stood in St. Peter's Square in Rome on the day when the pope comes out on the balcony and blesses the city and the world. Sure enough, there was Zichichi on the balcony with the pope! Then a man standing next to the visitor tapped him on the shoulder and asked, "Can you tell me, please, who is that up on the balcony with Zichichi?"

When we returned home, I walked over to my Harvard office to pick up mail and messages. In my mailbox, there was a handwritten note from Norman Ramsey, an expert on the sort of experiment being done to look for violations of space-inversion symmetry that would be shown by the rotation of polarization of light waves passing through bismuth vapor. Ramsey's note said that both the Oxford and Seattle groups were setting upper limits on this effect, which seemed to rule out the violation of space-inversion symmetry expected in the electroweak theory from the interaction of electrons with nucleons in the bismuth atom.

This disturbing news allowed three possible conclusions. First, the electroweak theory in its original form might be wrong. But it was hard to imagine any other theory that would lead to the same successes that had already been scored by the electroweak theory in accounting for the details of neutrino scattering. Next, the experiments might be wrong. But the groups at Oxford and Seattle were among the most highly regarded in this field. A little later, a group at Novosibirsk reported that they were seeing the rotation of polarization in bismuth vapor at the rate expected in the electroweak theory, but Russian experimental physics then did not have a good reputation in the West. Finally, the bismuth atom is pretty complicated, with its eighty-three electrons. My own guess at the time was that the electroweak theory was correct and the Oxford–Seattle experimental results were correct, but the atomic theory used to calculate the expected mixing of atomic states in the bismuth atom caused by the violation of space-inversion invariance in the interaction between electrons and the nucleus was just too difficult to have been done correctly. Time would tell, but for the present, no one could say that the electroweak theory was in the bank.

We left for Stanford at the end of the summer. We rented a nice house with a wonderful orchard. Its grand persimmon tree was to play a rather important part in my life.

I had suspected that Louise, who had not yet achieved her eventual celebrity in her fields, might have been invited just to secure a visit from me, although roundabout methods were not needed. But I have never felt so unattached to a host department. My recollections may be at fault, but as I recall, our only invitations that year came from the law faculty. Once Sid Drell and his wife stopped by to say hello on their way to buy ice cream. I had no contact with Wolfgang Panovsky. Fortunately, Helen Quinn was moving to Stanford, and I had the occasional happy benefit of her presence. Lenny Susskind, a good friend, did not join the Stanford Faculty until 1979. We had had many dear friends at Berkeley, so I could not blame this phenomenon on California. Today, I can connect it in my mind with the coldness I had felt when in early days I would attend a talk down at Cal Tech. I guess I was "that brassy, over-confident tyro." Of course they enjoyed taking me down a peg. And here I was again, now having some success to boot. Louise's invitation, like many others' was part of a broad visiting program to compensate for Stanford's isolation and the resultant absence of its traveling faculty.

But as I saw it, under the persistent influence of the *The Red Shoes*, we were all engaged together, and encouraging each other, in a grand joint enterprise dating from Galileo and coming down to us through Galileo, Newton, Maxwell, and Einstein – especially Einstein. And I celebrated the successes of comrades in this enterprise. So I could not fathom the distancing I felt that year at Stanford.

Whatever it was, it left me very much at liberty to do what I best loved, which was to work at home, the television on at the History Channel, and Louise in my life. Although electroweak theory was hung up on the seeming contradictions of experiment, in the early autumn, I put the final touches on *The First Three Minutes*, the book on cosmology that had been suggested to me in 1973 by Daniel Bell and Erwin Glikes. This was my first book for general readers. It was published in spring 1977 and received very good reviews, won a prize from the American Institute of Physics, and eventually appeared in twenty-three languages. (As this is written, I have just agreed to publication in a twenty-fourth language, Tamil.) When Glikes visited Palo Alto that spring, I asked him what had impelled him to offer me a book contract. He said that with my title, *The First Three Minutes*, and his ability to sell foreign rights, the book could not fail.

In my own mind, my cosmological turn had a lot to do with the paperback by Chandrasekhar that Louise, with all the enthusiasm of her freshman course in astronomy, had bought as a present for me when I lay in a hotel bedroom on Telegraph Avenue, newly arrived but immobilized with a slipped disc, in Berkeley, California. My early interest in cosmology also had to do with the 1964 discovery at Bell Labs, mentioned in Chapter 11, of a three-degree background radiation.

The thing about *The First Three Minutes* that most gratified me was the serious reading that astronomers gave this merely "popular" book. It seems to have brought cosmologists and particle physicists together. We could study the universe in our telescopes deploying the same fundamental physics with which we study the atom in our accelerators. To some extent, we had already begun to do so. Expatriates in a foreign country.

I had an invitation to contribute an article to a festschrift for Rabi, and took the occasion to follow up some ideas I had about quarks. The interactions among quarks are very strong in observed particles like protons, neutrons, pions, and K mesons, particles that are composed of quarks and antiquarks of the three lightest varieties. These interactions complicate our understanding of the quark masses. Nevertheless, there are theorems of current algebra that allow the use of the measured masses of observed particles to calculate ratios of light quark masses. One striking result I found was that the mass of the down-quark is about 1.8 times the mass of the up-quark. The reason that the neutron, with two down-quarks and one up-quark, has almost the same mass as the proton, with two up-quarks and one down-quark, is that the masses of the protons and neutrons arise almost entirely from the strong interactions, which do not distinguish between up- and down-quarks, and very little from the masses of the quarks they contain.

In March 1977, I learned that experimenters at Fermilab were seeing strange events in which a neutrino striking a nucleus turned not into a single muon, which is typical, but into three muons, something that would not happen in the original electroweak theory. I was willing to suppose that we did not yet have the true theory, but if a new theory was needed, I damned well wanted to be one of those who constructed it. I discussed this by telephone with Ben Lee, and he agreed to fly out to California to work on it with me.

Ben Lee's collaboration was a very special thing for me. Ben was one of the very few physicists who had absolute confidence that Salam and I had not merely furnished an example, but had described electroweak theory. I remember Ben and me, sitting at a patio table in our Palo Alto backyard, under the persimmon tree, laden with huge fruits, working out the details of an extended version of the original electroweak theory. Not long after our paper was published, we learned that the experiments at Fermilab had been wrong. I was reminded of a half-serious remark of the British astrophysicist Arthur Eddington: "One should never believe any experimental result until it has been confirmed by theory."

One particularly good thing came out of Ben's trip to California. He and I spent some time talking about cosmology, and developed an idea of how to use astronomical observations to set a lower bound on

FIGURE 12.1 Ben Lee in February 1977

the mass of any new kind of massive neutrino, left over from the early universe. We assumed that such a particle would be stable, but could annihilate in collisions with its antiparticle, as do ordinary neutrinos. The lower the neutrino mass, the slower the rate of annihilation, and so more of these particles would be left over from the early universe. Paradoxically, if these heavy neutrinos had a mass less than about 10–100 proton masses, there would be so many of them that their total mass density would exceed a bound set by the rate of expansion of the universe. Our calculations were correct, but in our presentation we missed an important opportunity. There already was some evidence that most of the matter of the universe is in some dark form, not composed of ordinary atoms or their constituents. The total mass density of the universe can be estimated from observations of the motions in clusters of galaxies and of the expansion of the universe, while it is possible to estimate the density of ordinary matter from the comparison of calculations of nucleosynthesis in the first three minutes with observation of the abundance of elements in interstellar matter that has not yet been cooked in stars.

The existence of cosmic dark matter became much better established later, when space-based and ground-based observations of inhomogeneities in the cosmic background of microwave radiation showed that the cosmic density of all matter is about six times the density of ordinary matter. The identification of the particles of dark matter has become a major preoccupation of particle physicists. The leading candidate has been "WIMPs," an acronym for weakly interacting massive particles. These are just like the massive neutrinos considered by Lee and me, and today calculations of their cosmic density follow the same lines as in our 1977 calculation. The fact that a mass of order 10–100 proton masses can give the observed dark matter density is often called the "WIMP miracle." I wish that, instead of writing about heavy neutrinos, we had referred to the more general possibility of WIMPs, and instead of giving a lower bound on their mass, we had emphasized that a WIMP with that mass might be the particle of dark matter.

I have done most of my best work without collaborators, but my work with Ben Lee had gone so well that I looked forward to collaborating with him again in future. He had become enthusiastic about working on cosmology. Shortly after returning to Cambridge from Stanford, at the beginning of the summer, I was shocked to learn that Ben had been killed in an automobile accident. A truck had crossed the meridian on a highway and smashed into Ben's car. It was a terrible loss for physics, and especially so for me. I was honored to give the summary talk at a physics conference in Ben's honor that was hastily called to be held at Fermilab in October. At the end of my talk, I remarked that Ben would have enjoyed the great variety of topics discussed at the conference, because throughout his working life he had tried to break through the barriers of specialization, and to cover the widest possible range of problems in physics. I should have added the more personal observation that Ben had a remarkable sweet personality, and so that knowing him had been a great pleasure.

Back in Cambridge, in the autumn of 1977, I started a term as a member of the Council of Harvard's Faculty of Arts and Sciences. I generally avoid service on university committees, but I agreed to serve on the Council because I was new at Harvard and it seemed like a good way to meet faculty in other departments. My other reason had to do with the room in which the Council met. The Faculty Council met in the large central room of University Hall, which was designed in 1813 by Charles Bulfinch, the architect of the golden-domed Massachusetts State House. This room had pastel-colored walls and moldings, was hung with paintings of Harvard worthies, and was filled with light from the many large windows. I felt I could sit in that Bulfinch room happily through the dullest committee meeting.

That autumn, I started worrying about some ideas I had heard about in the spring at Stanford from Helen Quinn, who had completed her move there from Harvard. She and her co-worker Roberto Peccei had been concerned about a problem raised by the 1976 work of 't Hooft I mentioned earlier. The photon-like gluon fields with which quarks interact can be in various twisted configurations, whose various contributions to physical processes are characterized by a real number, θ (theta). A nonzero value of θ would violate the invariance of the strong interactions under a reversal of the direction of the flow of time. Among other effects, this would produce an asymmetry in the electrically neutral neutron, with some density of positive charge at one end and an equal and opposite negative charge density at the other end. Experiments (initiated earlier by Norman Ramsey) had put such stringent limits on this asymmetry that one could conclude that θ has a magnitude less than one-billionth. Why so small? Of course, it could simply be assumed that the strong interactions respect invariance under time reversal so that θ has to be zero, but this would give up one of the attractive features of quantum chromodynamics, that time-reversal invariance is an automatic consequence of the simplicity of the theory imposed by the requirement of renormalizability. It was already known that time-reversal invariance is not respected by weak interactions, so it could not be supposed to be a fundamental symmetry of nature.

In March 1977, Peccei and Quinn proposed a theory in which θ is a dynamical variable that, like a ball rolling into a hole, evolves to a value $\theta = 0$ that minimizes the energy, in this case the energy of empty space. That autumn, it occurred to me that Peccei and Quinn, in effect, were assuming a new broken symmetry, something like the broken chiral symmetry underlying current algebra. This meant that these theories would imply the existence of a "Goldstone boson," a new spinless particle like the pion. But since the Peccei–Quinn symmetry, like chiral symmetry, is broken by the nonzero values of the quark masses, this new Goldstone boson, like the pion, would not be massless, but just very light.

The same ideas occurred at about the same time to Frank Wilczek. We agreed to call the new particle the "axion," and both sent papers about it to *Physical Review Letters* early in December 1977. The axion appearing in the original models studied by Wilczek and me has been ruled out by accelerator experiments, which could be interpreted to set a limit on their production in collisions, and by astronomical observations on the rate of cooling of red giant stars. But a number of theorists (including my Texas colleague Willy Fischler) have pointed out that it is still possible that axions exist if the Peccei–Quinn symmetry is broken at a sufficiently high mass scale, above about 10 billion proton masses. (The higher the symmetry breaking scale, the lighter the axion and the more weakly it interacts with ordinary matter.) Indeed, if the symmetry breaking scale is roughly a trillion proton masses, the axion could be the long-sought dark matter particle, instead of the WIMP. So axions continue to be the subject of concern for many physicists.

About a decade after the original axion papers were published, I received a package from physicists at a summer school, with a note saying, "We have found it!" Inside the package was a box of a detergent called Axion.

In the winter of 1977, I was awarded the Dannie Heineman Prize for Mathematical Physics by the American Physical Society jointly with the American Institute of Physics. It was a very welcome award because I have never considered myself especially strong mathematically.

FIGURE 12.2 The author receiving the Heineman Prize from Willy Fowler

In February 1978, I testified for the first time before a congressional committee. It was a joint hearing of the Subcommittee on Science, Technology, and Space of the Senate Committee on Commerce, Science, and Transport, and the Subcommittee on Science, Research and Technology of the House Committee on Science and Technology. Unlike my future congressional testimony, this time there was no controversial issue – the committee members simply wanted an update on open problems in physics. I gave a lightning nonmathematical talk on galaxy formation, turbulence, and the forces of nature. During this visit to Washington, I was invited over to the White House to be introduced to President Carter by his science advisor Frank Press, whom I had known at MIT. Frank was a geophysicist, who had made essential contributions to our knowledge of how to detect distant underground explosions, a necessary capability if we are ever to have a comprehensive nuclear test ban. Later, Frank was president of the National Academy of Science. Neither Carter nor I said anything memorable, but I have a nice photo of me meeting a president for the first time. I would meet him there again in November 1980.

A little later that spring, the attention of particle physicists became focused on an experiment under way at the Stanford Linear Accelerator Center (SLAC). A SLAC–Yale collaboration headed by Charles Prescott was studying the scattering of electrons by deuterons, the nuclei of heavy hydrogen. The new thing in the experiment was that the electrons were polarized – they could be selected to be spinning to the left or the right as they moved. If there was a difference in the scattering rate between left and right polarizations, this would show the kind of violation of space-inversion symmetry due to Z particle exchange that was expected in the electroweak theory, but apparently had been ruled out by the experiments with bismuth at Oxford and Seattle.

Late that spring, I was delighted to learn from a friend at SLAC that this experiment was showing just the level of left–right asymmetry that was expected in the electroweak theory. The report of this experiment published in July 1978 stated that the asymmetry agreed

FIGURE 12.3 Meeting President Carter at the White House

with the prediction of the electroweak theory of Salam and myself for a weak mixing angle close to 26°, the same as indicated by experiments on neutrino scattering. So there was nothing wrong with the electroweak theory. It turned out that there was nothing wrong either with the calculations that used the electroweak theory to calculate the violation of left–right symmetry in the bismuth atom. It was the Oxford and Seattle experimental results that were wrong.

A year or so later, I asked one of the members of the Oxford Group if they knew what had gone wrong. No, he said, and now they never would, because they had torn down their experimental setup to use its parts for a different experiment.

There was still a large remaining uncertainty in the electroweak theory. Is the breakdown of the underlying symmetry due to some field having a nonzero value in empty space, as Salam and I had assumed, or is it a dynamical symmetry breaking, due to some new strong interaction, as in the breaking of the symmetry underlying current algebra in a 1961 model of Nambu and Jona-Lasinio. In 1976, I had explored the latter possibility for a general class of gauge theories of the Yang–Mills sort, without reference to any specific model. Then in the spring of 1978, I realized that, in any version of the electroweak theory with dynamical symmetry breaking, one would get the same successful formulas for the W and Z particle masses as in the original electroweak theory.

The same result was independently found at about the same time by Lenny Susskind. By analogy with the color forces that lead to the breakdown of chiral symmetry in quantum chromodynamics, Susskind invented the name "technicolor" for the forces responsible for the breakdown of the electroweak symmetry.

Technicolor had the very attractive feature of offering a possible partial solution for the hierarchy problem, the mystery of the huge disparity between the mass scale of electroweak symmetry breaking and the vastly higher scales at which Georgi, Quinn, and I had found that the strong and electroweak interactions couplings come together and the scale at which the strength of gravitation is similar to the strength of other forces. The technicolor coupling might, like the color coupling, become strong slowly with decreasing energy, becoming strong enough to break the electroweak symmetry a little sooner than the color coupling becomes strong enough to give the proton its mass. But although technicolor could do a good job of accounting for the W and Z masses, it was not clear how quarks and leptons could get their masses. Nevertheless, technicolor has remained for decades in theorists' minds as a candidate for electroweak symmetry breaking.

In the spring of 1978, for the first time I received invitations to receive honorary doctoral degrees. The first was from Knox College, a small liberal arts college at Galesburg on the Illinois prairie. Galesburg was one of the sites of the Lincoln–Douglas debates in the 1858 campaign for a seat in the U.S. Senate. At Knox College, I was shown a large window through which Lincoln had passed on his way out from the main college building to the debaters' stage, saying, "At last I have gone through college."

The other degree I received in 1978 was from the University of Chicago. I especially prized this honor, because Chicago's rules allow honorary degrees to be awarded only for excellence in academic disciplines. No one was eligible as a donor, movie star, or sports hero.

For some time, I had wanted to visit China. For Americans in the 1970s it was a mysterious place, cut off from the West by a strange ideology. A delegation of scientists headed by Murph Goldberger had gone to China shortly after Richard Nixon had famously done so in the early 1970s, but there were no international physics conferences in China like the one that had brought me to Russia in 1964. I cannily figured that, instead of writing to Chinese physicists proposing a visit, I would be treated better if the invitation were initiated by them. So I hinted, I thought subtly, to Tsung-Dao Lee at Columbia that I had often thought I would like to visit China some day. Sure enough, I soon received an invitation from the Chinese Academy of Science. To my chagrin, it began, "We have heard of your urgent desire to visit our country." As it happened, I did also have an invitation to give the rapporteur's talk on weak interactions at the 19th International Conference on High-energy Physics, to be held in Tokyo on August 23–30, 1978, so I planned to take Louise and Elizabeth to Tokyo for that week, and then on to China.

The thrust of my talk was that everything being observed was working out as expected theoretically. While in Tokyo, I checked our flight reservations to Beijing, and found that for some unknown reason American Airlines had canceled our reservations. And they had no seats available. In order to get to Beijing on time, I had to book first-class seats for us on Iran Air. It was expensive, but had its compensations. In first class on Iran Air (still in the days of the Shah), we had the best caviar of our lives, pale gray and meltingly flavorful, served in abundant heaps, with refills.

When our plane landed in Beijing, we were immediately put in a small gray car waiting on the tarmac. Salam was visiting Beijing at the same time, and had been assigned a big black car. Also, while we had only a radio in our hotel room, he had a television set. Amusingly, in a classless society like China, strict rules assign people to classes anyway. Salam was director of an institute, the International Centre for Theoretical Physics in Trieste, and I was not. Indeed, Salam's Institute has grown enormously in size and stature and I regret that Abdus, a dear friend, is not alive to see it.

Judging from the variety of costume and language, the crowd in the bar of our hotel was extraordinarily cosmopolitan. One evening, we met Édouard Brézin of the University of Paris, a leader in the study of critical phenomena whom we had known and liked when he visited the Harvard physics department a few years earlier. Mark Twain said of San Francisco that sooner or later everyone is seen there. Now it is Beijing.

We arrived home in Cambridge. That year, the world was commemorating the 100th anniversary of Einstein's birth in 1879. There were meetings at the American Academy in Boston and the American Philosophical Society in Philadelphia, at which I spoke.

In the autumn of 1978, I had been asked by Stephen Hawking to contribute an article to a book he and Werner Israel were assembling for the Einstein centennial. I planned to review efforts to deal with the notorious infinities that appear in calculations in the quantum theory of gravitation. The principle of equivalence of gravitation and inertia on which general relativity is based does not allow any theory of gravitation in which infinities can be removed by a renormalization of physical quantities, as in the Standard Model.

Then I had an idea. It was well known that, in any sort of quantum field theory, the various quantities known as couplings that determine the strength of interactions depend on the energy at which they are measured. I had learned from my reading on critical phenomena that these phenomena, including some phase transitions, depend on the existence of a fixed point – a point in the mathematical space of all the couplings at which the energy dependence of the couplings disappears. Typically, as the energy decreases (or the scale of distances increases), the couplings are attracted to the fixed point – provided that they satisfy a finite number of conditions. In that case, the system exhibits order over large distances.

For instance, as I already mentioned, the phase transition in which the spins in a ferromagnet line up over long distances requires a single condition, that the temperature should take a certain value. We can think of these conditions on the couplings as a requirement that, near the fixed point, the little vector that separates the couplings from their value at the fixed point has no component along a certain critical surface of finite dimensionality. This is because the number of conditions equals the number of components that have to vanish, and this number equals the dimensionality of this surface. In the case of ferromagnetism, this surface is a one-dimensional curve, and so there is just a single condition. If the vector separating the couplings from the fixed point has any components on such a surface, then these components do not vanish at low energy or large distance, but typically blow up.

Running the argument backwards, it follows that, if near a fixed point the little vector separating the couplings from the fixed point has components only on the finite-dimensional critical surface, then these couplings are attracted to the fixed point as the energy increases.

For example, in quantum chromodynamics, as long as we stick to the simplest renormalizable theory with a single coupling, the coupling vanishes as the energy goes to infinity. In this case, the theory is asymptotically free – that is, the fixed point is at zero coupling, where the theory becomes a theory of free particles. Adding non-renormalizable terms takes the theory off the critical surface, and it is no longer asymptotically free.

More generally, the fixed point will not be at zero coupling, but there will still be finite-dimensional critical surface, and hence a theory with a finite number of couplings that approach the fixed point, and therefore do not blow up as the energy goes to infinity. I called such theories asymptotically safe. Though not renormalizable in the same sense as quantum chromodynamics or the electroweak theory, they share the important feature of renormalizable theories that a condition of good behavior at high energy limits the free parameters of the theory to a finite number.

In my article in the Einstein centennial volume, I proposed that gravitation is governed by an asymptotically safe theory. That's easy to say, but with a fixed point nowhere near zero, this can't be shown using the usual techniques of calculating the rate of change of the couplings as a series in powers of the couplings. I did what I could to make this proposal plausible by considering general relativity with the number four of space-time dimensions replaced with a continuous variable. (This was a trick already well known in quantum field theory, exploited especially by Schwinger and 't Hooft.) I was able to show that, in $2 + \varepsilon$ space-time dimensions (with ε being an infinitesimal positive number), there is an asymptotically safe theory with a single nonzero coupling. At the fixed point, this coupling is of order ε, so the energy dependence of the coupling can be calculated as a series in powers of ε, confirming the asymptotic safety of the theory.

Of course space-time in the real world is four-dimensional, so $2 + \varepsilon = 4$, and $\varepsilon = 2$. Each additional term in a series in powers of 2 is likely to be bigger than the terms that went before, so this gets one nowhere. It is therefore difficult to show that there is an asymptotically safe theory of gravitation in the real world. This problem has been addressed by a number of theorists, chiefly in Germany and Italy, so far with results that are encouraging but not conclusive.

In the autumn of 1978, Louise, who was beginning to make a name for herself, received an invitation to teach at the University of Texas Law School in Austin in the 1979–80 academic year. Unlike the year at Stanford, this would be a lookover visit, leading to the offer of a faculty position if all went well. Louise explained that she had recently spent a year away from home at Stanford, and had dragged the whole family along with her. She did not feel ready for another year-long visit anywhere. The dean kindly suggested that Louise might come to Austin to teach in just the 1979 summer session, and this she accepted.

In the spring of 1979, I gave talks in honor of the nuclear physicist Amos De-Shalit at the Weizmann Institute in Israel in March, and in honor of Hans Bethe at Cornell in April. In June, I gave the closing talk, "Beyond the First Three Minutes," at a conference on the early universe at Copenhagen.

At commencement time in 1979, I was awarded honorary degrees at Rochester and Yale. The Yale commencement was memorable for the brief late appearance of the great premier danseur noble Mikhail Baryshnikov. With me and all the other honorands already waiting patiently on the stage, Baryshnikov arrived just in time for his own award. He marched down the center aisle with his cape flowing behind him, received his degree, and immediately marched back out. I was reminded of a story I had heard at Harvard about the eminent choreographer George Balanchine, who was once told by his colleague Lincoln Kirstein that he, Balanchine, was to be awarded an honorary degree by Harvard, but would have to attend the June commencement personally. Kirstein recalled that Balanchine had asked, in his thick Russian accent, "Vhat ees dees Hyarvard? Dohn dey knaw I hyev rehyarsal?"

In Austin, we rented the house of a noted chemist, Esmond Snell. I had no visiting appointment at the physics department, but I gave a talk there in honor of Albert Schild, who had just passed away. He had founded the excellent group in Austin doing research on general relativity.

Austin was then a much smaller, much quieter, town than now. It was the state capitol and a college town, without the high-tech companies that moved in later, and without the skyline of tall buildings now clustered there. It was full of little white houses as far as the eye could see, but it had a little downtown with a Woolworth's and Scarborough's, a department store. Across Town Lake, there were the offices of its newspaper, the *Austin American-Statesman*. The capitol building was, and is, western in spirit and truly beautiful.

Austin also could boast a longstanding tradition of country music writing and performance. It had true celebrities of the genre, like Willie Nelson and Jerry-Jeff Walker. It had characterful honky-tonk joints like The Broken Spoke. It had a TV show showing off its special country styles, *Austin City Limits*. It would come to rival Nashville for country music. On a downtown corner, there was a large open area with benches, tables, a bar, and a stage, calling itself the Armadillo World Headquarters. For the price of a beer, you could sit and listen to one country group voluntarily succeeding another every twenty minutes. When it was destroyed a few years later to make room for a tall building, the earthen ground was found littered with thousands of marijuana cigarette butts.

At the Law School, Louise was introduced to a young faculty member, Philip Bobbitt, who would become one of our dearest lifetime friends. Most people we knew in Austin had moved there from elsewhere, often from other towns in Texas, but Philip had been born and grew up in Austin. Philip was educated at Princeton and the Yale Law School, but had kept a stately southern voice and manner. He was a nephew of Lyndon Johnson, and introduced us to his "Aunt Bird." He was then a professor at the Law School, and we were in Austin when Philip purchased his grand wood-paneled old house on a large lot in Old Enfield, the toniest residential area in Austin. He once told us that when the Law School was trying to recruit new faculty, the dean would arrange for them to visit Philip's house, remarking that this was the sort of house in which assistant professors lived in Austin, so one could imagine how full professors lived.

Shortly after Louise's summer visit, the University of Texas School of Law offered Louise a full professorship, to start in the autumn of 1980. For an unusual reason, the Texas offer was irresistible. It was commonly understood that the University of Texas School of Law was and remains one of the greatest law schools in the nation. Today's reader may not know that its greatness had been celebrated by the Supreme Court of the United States, that every element of its greatness had been there described, and that every law professor and every law student in the country was reading that Supreme Court opinion. The case was the early civil rights case Sweatt v. Painter. At the time of this case, the University of Texas Law School had been racially segregated. An African-American plaintiff, Heman Sweatt, sued to be allowed to attend the same classes in the same classrooms by the same professors as white students. In those days, separate accommodations for black people had been held not to deny them equal protection of the laws, as long as the separate accommodations were equal to those provided for whites. The University argued that its separate accommodations for black students were equal to those for whites.

But the Supreme Court used the case to move the country forward, pointing out that the greatness of the University of Texas School of Law could not be equaled by special arrangements. There was no equal to its great law library, the fifth largest in the world, its distinguished faculty of famous scholars, its handsome interiors with its finely appointed moot courtroom, its prestigious *Law Review*, its luxurious lounges, its gold-framed portraits, and its proud traditions. These glories and more all did exist at the time, and Louise loved them all. There could be no equal to any of this.

Never was there anything in our lives less likely than a move that would separate us or anything less likely than a move of one of us to Texas. How could one persuade a New Yorker enjoying the sophistication of Cambridge to move to Texas? At about that time, a Texas magazine published an issue with a good-humored cover. It depicted a mean old sheriff in a big hat, boots crossed up on his desk. What could be more off-putting? Except that this mean old sheriff had a little puppy on his lap, and was petting it. What could be more reassuring? A stereotype was good-naturedly blown away.

The Supreme Court's salute to the University of Texas School of Law, at the time in every American law student's casebook, had the effect of blowing away stereotypical thinking. The offer was impossible to refuse in any case, but the Supreme Court's imprimatur put the seal on it.

Some of the glories of the place, so glowingly described by the Supreme Court, have been dismantled by time, fate, and unthinking modernizing or economizing deans. But it remains a very great law school indeed, and one that came to appreciate Louise's genius as much as she appreciates the Law School.

For us, separation was the hardest part of this development. I could not imagine permanent separation from Louise. It was almost as hard to imagine leaving Harvard. As Higgins Professor at Harvard, I had the top academic job in American physics. We had both enjoyed our summer in Austin, but I loved being at Harvard. I certainly did not want to initiate discussions with the Texas physics department. We managed to persuade ourselves that, with Elizabeth at Andover, we could teach in separate distant cities, with me making frequent visits to Austin, and all of us getting back to our beloved house in Cambridge for Christmas, and for summer vacations. That was the plan. Our hearts would be together in Cambridge, however separated we ourselves had to be for brief thirteen-week stints of teaching.

At some time in 1979, I received an invitation to contribute to a festschrift celebrating Julian Schwinger's 60th birthday in 1978. As I mentioned previously, my work in 1967 on the theory of low-energy pion interactions had benefited from Schwinger's advice. He had urged me to construct a theory of low-energy pions directly, on the basis of invariance under the chiral symmetry transformations that would apply in a theory of only pions. I had worked this out in the summer of 1967, keeping only those terms in the theory's equations that, in the simplest approximation, would yield results for scattering rates that are proportional to the minimum number of powers of energy. Calculations in this theory in the simplest approximation automatically gave the same finite results as current algebra for processes like the scattering of low-energy pions by pions.

Now, in preparing my article for the Schwinger festschrift, I played with the idea of calculating physical quantities without limiting these calculations to the simplest approximation. I knew that this would give results as a series in powers of the energy. The term with the smallest number of powers of energy would be the same as I had already calculated, and of course would dominate the results for very low energy. But there would be an infinite number of terms proportional to higher and higher powers of energy. I also knew that these higher power terms would be riddled with infinities. But suppose I did not limit myself to the original theory, but instead included in the theory all of the infinite number of possible terms in the equations of the theory that are constrained only by symmetry principles. After all, why should a theory of a particle like a pion, which is a mere quark–antiquark composite, have any special simplicity? I found that, for each infinite term arising in these calculations, there is always a term in the equations of the theory proportional to a constant that can be supposed to cancel the infinity, just as a redefinition of the electron's mass and charge cancels the infinities appearing in quantum electrodynamics. Furthermore, the terms of each order in energy involve only a finite number of the infinite number of constants in the theory, so this theory can actually be used to calculate any finite number of terms in a series of powers of energy, with only a finite number of constants needed to be taken from experiment.

But this analysis breaks down if the energy is too large. How large is too large? The numerical value of any energy depends on what units one uses: calories, electron volts, kilowatt hours, whatever. The probabilities for physical processes cannot depend on an arbitrary choice of units, so a series in powers of energy that we use in calculating probabilities must really be a series in powers of energy divided by some quantity W, with the same units as energy, such that the ratio does not depend on the choice of units. But the series is then useful only if the energy is much less than W, in which case this ratio is much less than 1, and high powers of the ratio are much smaller than low powers. Inspection of the first terms in the theory of low-energy pions suggests that, in this theory, W is about 1.2 billion electron volts. For pion energies approaching a billion electron volts (a high energy by the standards of nuclear physics, but much less than the energies accessible at today's accelerators), these calculations lose all value, and one has to know something about the underlying theory, which for pions is quantum chromodynamics.

This theory served as a template for what has come to be called effective field theory. My Texas colleague Willy Fischler says I am the father of effective field theory, and I like to think so. The key idea is that quantum field theory is just a way of implementing various symmetry principles and the general principles of quantum mechanics. As long as we include every term in the equations of a theory that is not forbidden by symmetry principles and quantum mechanics, in using the theory, we are not making any assumptions beyond symmetry principles and quantum mechanics. But even if the theory cannot be wrong, it may not be useful. It could be of use only at energies much less than some characteristic energy W.

It has become widely, though not universally, accepted that general relativity, in the form presented by Einstein in 1915, is just the leading term in an effective field theory of gravitation. Only in this case the energy scale W, at which the low-energy approximation breaks down, is not a measly billion electron volts, but is the Planck scale, 10^{18} billion electron volts, that I mentioned earlier. From this point of view, there is no incompatibility between gravitation and quantum mechanics, as has often been thought. The real problem is that we do not know what nature is like at the Planck scale.

Encouraged by the new perspective of effective field theory, I began to think about the sort of new terms that might be added to the equations of the original Standard Model, and their physical effects. This might seem like a dumb idea. The effects of these new terms would be suppressed by powers of the ratio E/W, where E is the energy of the process under consideration and W is the energy of some underlying theory, perhaps the energy 10^{15} GeV, where the strong and electroweak couplings come together, or the Planck energy 10^{18} GeV. At any energy E that is available in accelerator laboratories, the ratio E/W is incredibly small, so how can we ever hope to detect physical effects proportional to powers of this ratio?

The only hope is to look for the violation of conservation laws that would otherwise be respected. The equations of the original Standard Model are so constrained by exact symmetries like charge and color conservation, and by the condition of renormalizability, that they cannot violate the conservation of quantities known as baryon and lepton number.

Baryon number is the number of protons and neutrons and certain heavier particles that decay into protons and neutrons, minus the number of their antiparticles. (Since these particles consist of three quarks, baryon number can also be defined as one-third the number of quarks minus the number of antiquarks.)

Lepton number is the number of electrons, and certain heavier particles like muons that decay into electrons, plus the number of neutrinos, minus the number of all their antiparticles.

It is these conservation laws that keep ordinary matter stable and that, for instance, forbid the decay of a proton into a positron (the electron's antiparticle) and a photon.

I saw no reason why additional non-renormalizable terms that might be added to the equations of the Standard Model should be expected to respect the conservation of baryon or lepton number.

I set out to catalog the terms in the field equations that respect all known exact symmetries and are suppressed by a minimum number of factors of E/W that would violate the conservation of baryon and lepton number. I found terms of order $(E/W)^2$ that would produce a very slow but perhaps detectable rate of proton decay. The same results were found independently at the same time by Frank Wilczek and Tony Zee. I also found a term suppressed by only one factor of E/W that violates the lepton number but not the baryon number, and would produce very small neutrino masses, less than a millionth the mass of the electron.

There already were hints of such small neutrino masses. For years, starting in 1970, Ray Davis monitored a tank of 100,000 gallons of cleaning fluid in a gold mine in Lead, South Dakota, looking for the sudden appearance of electrons produced by neutrinos from the Sun striking chlorine nuclei in the molecules of the fluid. Elaborate calculations of the production of neutrinos in the Sun's core by John Bahcall gave a definite prediction of the rate at which these events should be seen. As time passed, it became clear that the number of electron appearances was about a third of the result expected from Bahcall's calculations. A possible explanation was provided by an earlier speculation of Bruno Pontecorvo, that neutrinos have mass, but the states of definite mass are mixtures of the states produced or destroyed along with electrons or muons or the heavier tau leptons. The neutrinos produced in the Sun are initially of electron type, but they are superpositions of three different mass states, which by the time they reach South Dakota are all mixed up with each other, only about a third being of the type that could turn into electrons in the cleaning fluid.

By now, there are several terrestrial experiments that verify this explanation. Neutrinos do have mass, less than about a millionth of the electron mass, but the precise values are not yet known.

In October 1979, we had further happy news. I was to be awarded the 1979 Nobel Prize in Physics, to be shared with Abdus Salam and Shelly Glashow. The news did not come to me from a trans-Atlantic telephone call, but from a call from my father, who had heard the news on CBS radio. He was of course very happy about it, and I was happy that the prize had come early enough so that he would know about it, though I grieved that my mother was long gone. Later I did receive notice of the prize in a telegram from the Royal Swedish Academy of Sciences, with a request to indicate my acceptance by telegram. I called Western Union in Boston. When I dictated to an operator a telegram accepting the Nobel Prize, she exclaimed in a thick Irish brogue, "Oh, that's marvelous, so you'll be meeting Mother Teresa!" It had already been announced that Mother Teresa would receive the Nobel Peace Prize that year, but I never met her – the Peace Prize is awarded in Oslo, not Stockholm.

On the day when the Nobel Prize was announced, Louise gave me some very wise advice: "Now you have to write some unimportant papers." I knew what she meant. Even when scientists who receive the Nobel Prize resist the temptation to become panjandrums like Heisenberg, issuing judgments about what others should be doing, there is a tendency for laureates to feel that one should stop doing the ordinary hard work of science, and instead go only for the next Big Thing. I take pride in the fact that I have continued working hard on minor problems and have written a large number – literally hundreds – of unimportant papers, both before and after the Nobel Prize. Whatever other physicists may have learned from these papers, I have learned a lot. And at least I have escaped panjandrum-itis.

I almost wish that the Nobel Prize had been more of a surprise. In 1976, Salam and I had been tipped off by the article I mentioned in *Dagens Nyheter* that we were under consideration. I also heard (improperly, I suppose) from a few laureates that they had nominated me. I understood that the prize was in abeyance while the experiments at Oxford and Seattle were casting doubt on the validity of the original electroweak theory, but confidence in the validity of the theory had been restored by the SLAC–Yale polarized electron scattering experiment in 1978.

We had two months of busy preparation before leaving for Stockholm in December.

I had to prepare the talk that I would have to give at the Swedish Academy of Science. My friend and Harvard colleague, the physicist-historian Gerald Holton, warned me that I should take this task very seriously indeed, because the Nobel lecture would in future be closely examined by historians of science like him. In order to find a quiet place where I could work on my talk without disturbance, Louise and I drove to a favorite country hotel, the Fox Hollow Inn in the Berkshire Hills of Western Massachusetts. Fox Hollow had been the summer home of the Westinghouse family. It sat on an extensive patch of rolling countryside. We had stayed there during many earlier summers. At Fox Hollow, one came down to breakfast, either on the solarium porch, or in the well-stocked library (it had once been a girls' school). We could swim and ride horses and watch other guests playing tennis. But this time, I would be working hard on my talk.

Finally, my talk was written and Louise, Elizabeth, and I were wardrobed and ready to leave for Stockholm. For the first time in our lives, arriving at a foreign airport, we did not have to stand in line to go through customs and passport control. In Stockholm, our passports were taken from us and we were guided to a VIP lounge. This was the way to travel! Because Sweden is an egalitarian society, this room was not labeled as a VIP lounge, but as an emergency medical station. There we met a young man from the Swedish foreign office and his wife, who would be our guides and minders during our time in Stockholm. After officials had done what they needed to do with our passports and luggage, we were put in a car that would be assigned to us. At the Grand Hotel, we were taken to a two-bedroom suite with a view of the Royal Palace in the Old Town island across the waters of Stockholms ström.

The Nobel Prize ceremony always takes place on December 10, the anniversary of Alfred Nobel's death. Two days before that, I was scheduled for my one big responsibility in Stockholm, to deliver a physics talk at the Royal Swedish Academy of Sciences, which selects the prize winners in physics and chemistry. Here is the beginning of my talk:

> Our task in physics is to see things simply, to understand a great many complicated things in a unified way, in terms of a few simple principles. At times our efforts are illuminated by a brilliant experiment, such as the 1973 discovery of neutral current neutrino reactions. But even in the dark times between experimental breakthroughs, there always continues a steady evolution of theoretical ideas, leading almost imperceptibly to a change in previous beliefs.

I went on to explain how the details of the Standard Model (aside from the values of numerical constants) can be seen as a consequence of the symmetries of the theory, together with a requirement of simplicity that is necessary for infinities to be eliminated by renormalization of a limited number of numerical constants, including masses and couplings. I ended with an explanation of effective field theory, remarking that, if the Standard Model is a low-energy limit of a more fundamental theory of enormously more heavy particles, then we should expect its equations to be corrected with very small non-renormalizable terms, suppressed by negative powers of the enormous masses or equivalent energies. I remarked that, since the conservation of baryon and lepton numbers in the Standard Model is an automatic consequence of its exact symmetries and renormalizability, one could expect that the very small non-renormalizable terms would violate the conservation of baryon and lepton numbers, leading for instance to proton decay and neutrino masses.

This seemed like a new idea to some in the audience. After my talk, one older physicist complained that I seemed to be saying that the fact that baryon and lepton conservations are automatic in the Standard Model suggests that they may not be really exact. That was indeed what I was saying.

During the prize ceremony in the town hall two days later, I was on the stage with the other new laureates and the Swedish king and queen, while Louise and Elizabeth in their lovely gowns had good seats near the front. All eyes were on Abdus Salam, who had opted to wear his national dress instead of white tie and tails. He was sporting a turban, and slippers with long curled toes, as in an illustration from the Rubaiyat. The ceremony proceeded according to a fixed plan. All I had to do when prompted was to walk to the king, bow, take the prize scroll, bow again, and go back to my seat.

That night, we dined at the customary Town Hall banquet. After entering, one descends a grand staircase to the banqueting level, serenaded by young men and women in their white student caps. Louise in her high heels feared that she might not make it safely down the stairs, but she was ably escorted by Prince Bertil, and had no problem. She told me that, with an incredibly strong grasp of her elbow, Bertil flew her down, about three inches above ground level.

The main course at dinner was reindeer with cloudberries, which I welcomed as an unfamiliar treat, not knowing that I would be fed a good deal of reindeer in the next few days. One of the three physics laureates had to make a brief ceremonial speech at the banquet. Shelly and I agreed to let Abdus do it. In his exotic costume, he would be more colorful and memorable than either of us could be.

FIGURE 12.4 The author receiving his Nobel Prize from King Carl Gustaf of Sweden

FIGURE 12.5 Glashow, Salam, and Weinberg at the Nobel Prize ceremony

The next evening, there was a ball at the Royal Palace, attended by laureates, members of Swedish academies and the Nobel Foundation, the diplomatic corps, and the royal family. It was fun after the ball to sit down for drinks with Swedish university students and sing beautiful Swedish Christmas carols – the strange syllables phonetically arranged for us foreigners.

December is a good time to be in Stockholm. The night comes in the middle of the afternoon, and store windows are bright with Christmas decorations. In the Old Town, there are sidewalk stands that sell hot glögg, a fortified sweet wine. In the long winter night, one takes breakfast before sunrise. So it is natural that the Swedes celebrate the day of returning light, Saint Lucy's day, December 13. Though the solstice is still a week off, John Donne had called this day "the year's midnight."

It was supposed to be a surprise, but we knew that, on the morning of that day at the Grand Hotel, a train of maidens carrying candles and singing "Sankta Lucia" would customarily come into the room of each of that year's laureates and serve coffee and saffron buns. Louise, Elizabeth, and I had woken up early to be ready for this treat, and sure enough we heard girls singing "Sankta Lucia" approaching our suite. Then the sound fell off. We speculated – have they forgotten us? After a little while, I went to our door and peeped out. Here they were, coming down the hall! We all jumped back into bed, the door opened, and a dozen blonde girls came in singing, the lead girl with candles in her headdress. We exclaimed, "What a surprise!" Elizabeth told us they sang also to her in her own room. A good end to a good week.

Laureates are welcome to return to Stockholm for future Nobel Weeks. Some laureates do return now and then, including Glenn Seaborg, who was there at the ceremonies in 1979. But I never wanted to show up without a specific invitation. Louise and I returned to Stockholm for Nobel Week three times, when I did receive invitations. In 1991 and 2001, the Nobel Foundation was commemorating the 100th and 110th anniversaries of the first Nobel Prizes, and all laureates were invited back. In 1999, I served on a panel that had been convened by the Swedish Academy of Sciences to advise them on a possible prize to Gerard 't Hooft and his teacher and sometime collaborator Martinus Veltman. I had been strongly in favor of honoring 't Hooft for his work on the renormalizability of the electroweak theory, and vigorously supported his nomination. When it was announced that 't Hooft and Veltman were to receive the prize in physics for 1999, the panel was invited to Stockholm for that occasion. There was no secret about why the panel members were there, so the usual anonymity of the advisory process was in part suspended.

At a cocktail reception at the Academy of Sciences, Veltman confronted me, and complained that I had always opposed his recognition. I replied that I had been more enthusiastic about 't Hooft, but from the fact that we were both there in Stockholm he could judge for himself whether I had opposed his award.

By 1999, we would be living in Texas, so it was nice at the Nobel Ball to meet a fellow Texan, the US ambassador to Sweden Lyndon Olson, a large man with a jolly expression. We have seen Lyndon from time to time since then, always with pleasure. Another good thing was hearing a performance of Swedish Christmas carols by high school students in a small lobby in the Royal Opera House. Playing the recording we bought in Stockholm, of "Bethlehems Sterna" and other Swedish carols, never fails to evoke nostalgia.

On our way back from Stockholm to Cambridge in 1979, we visited physicists in Copenhagen, Zurich, and Geneva. It was a special pleasure in Geneva to have dinner with our dear old friend Bruno Zumino and his second wife, the physicist Mary Gaillard.

13 Gone to Texas

Back at Harvard, in the spring of 1980, I began to interact strongly with a remarkable young theorist, Edward Witten. All theoretical physicists have to be comfortable with mathematics and, as the world goes, are accounted good mathematicians. But the great majority of physicists, including me, do not have the sort of capability for incisive abstract mathematical thought possessed by real mathematicians. Pauli and Wigner may have had this mathematical strength, but probably not Einstein or Bohr.

Witten does. It can be seen in his work on general relativity, for which in 1990 he was awarded the highest prize in mathematics, the Fields Medal. After Witten received his PhD from Princeton in 1976, he applied to us at Harvard for a postdoctoral position, with a letter of recommendation from Murph Goldberger consisting of a single sentence, which as near as I can remember said, "If you do not hire him you are crazy."

Our work in 1980 explored what sorts of massless particles can exist. The known massless particles are photons and gluons, with spin one (that is, with an angular momentum equal to the Planck constant of quantum mechanics), and the graviton, the quantum of gravitational radiation with spin two. (The spin ½ neutrinos were then thought to be massless, but, as I mentioned, are now known to have small masses.)

Witten and I were able to show that, under very general assumptions (which are satisfied for instance in relativistic quantum field theories like the Standard Model), such theories do not involve gravitational fields, having neither a graviton nor any massless particle with spin higher than one. Thus, they cannot exist, even as a composite of other particles. If you want a theory of gravity, you have to insert the gravitational field yourself, and not hope to find gravitons as composites of other particles.

In addition to the pleasure of working with Witten, I reaped an extra benefit from this work: I always insist that authors of papers on which I collaborate should be listed in alphabetical order, so I am usually listed last. But not this time.

At some time in the 1979–80 academic year, I was invited to join the Saturday Club of Boston. This Club was founded in 1856. Its early members included Richard Henry Dana (one of the founders of Louise's old law firm), Ralph Waldo Emerson, Oliver Wendell Holmes, Henry Wadsworth Longfellow, Charles Sumner, and other New England notables. It was a famous tradition of the Saturday Club that Emerson would not attend the April meeting. From his home in Cambridge, he had sent a message: "I will not go to town while the lilacs are in bloom."

The Saturday Club represents Boston as it likes to think of itself as the Athens of America. It meets for lunch and conversation on one Saturday in each month. For many years, meetings were held at the Parker House Hotel. In my time, we met at the Union Club, overlooking Park Street and the Boston Common. When I joined the Saturday Club, its members included my friend Tom Adams, Louis Cabot, Archibald Cox, John Kenneth Galbraith, the *Boston Globe* editor Tom Winship, and Louise's former boss Charles Wyzanski. Judge Wyzanski was then, perhaps, the only Jewish member before my coming. From the beginning, there had been no women members, but soon after I joined it was decided to liberate the Club and offer membership to the new president of Wellesley College, Nannerl Keohane. If anyone had thought that this would change the character of the Club's conversations, Nan proved them wrong. Everything was exactly the same.

I was the youngest member of the Club, and the one with the least established ties to Boston. (I suspect that it was Tom Adams who put me up for membership.) But I like to think that I am clubbable, at least in the sense that I have no genes for shyness that would keep me out of conversations with anyone. Most members were friendly, but I found Cox arrogant and Galbraith argumentative. Once, when I was arguing for government spending on accelerator laboratories, my pet project, Galbraith made the common assumption that this would strip funds from programs of health and education. I reassured him that I had read *The Affluent Society*, and agreed with him that our economy should shift to more support for public goods, health, and education along with scientific research.

At one lunch, a strange thing happened. There was some sort of demonstration outside on Park Street. When a club member said that we had better not close the window shutters, there was a great laugh. After lunch, I asked what that was about. I was told that, on one Saturday during the Civil War, the black soldiers of the 54th Massachusetts Infantry Regiment happened to march past. A member who did not want to see black soldiers on the streets of Boston moved to close the shutters, and others strongly objected. The argument almost ended the Club. At any rate, now the members of the Saturday Club are regularly reminded of the 54th Massachusetts, because nearby at the corner of Park and Beacon Streets there is the great Saint-Gaudens bas-relief of the 54th marching past. The unforgettable film, *Glory*, at which Louise and I wept, well memorializes the 54th Massachusetts.

In the commencement season of 1980, I received an honorary doctoral degree from the City University of New York at their Staten Island campus. But in this season, I was on the giving as well as the receiving end. My term on the Faculty Council at Harvard was coming to an end, and I was put on the faculty committee to advise on nominations for Harvard honorary degrees. I was glad to see the nomination of the Mexican poet Octavio Paz, whom I knew slightly, having met him and his pretty wife Marie José – I think at one of the Kirchners' dinner parties. Paz was to become a rather close friend, hosting us at their apartment in Mexico City and visiting us when in this country. He was selected for an honorary degree, along with the great civil rights leader Bayard Rustin, and a few others.

At the commencement, I was honored to be Rustin's guide and minder. Since Texas was now much on my mind, I asked Rustin how he felt about Lyndon Johnson's role in advancing civil rights. Rustin told me that once, in the Oval Office, he had asked Johnson how it was that after years in the Senate opposing civil rights as a conservative Texas Democrat, when he became president, he did more for civil rights than any president since Lincoln. Rustin said that Johnson went to the window, seemed to reflect for a moment, and then answered, in the words of Martin Luther King Jr., "Free at last, free at last, thank God almighty, I'm free at last."

At the end of the summer, Elizabeth returned to Andover, and I flew with Louise to Austin and helped her move into the Snells' house, which we were renting for her again. Then I returned to our empty house in Cambridge. Loneliness settled in. I began to doubt whether separation could be endured for long.

I spent the 1980–81 and 1981–82 academic years at Harvard while Louise was teaching at the Law School in Austin. I would fly out to Austin to spend weekends with her. We threw some parties for the Law School crowd. When I visited, if nobody else was throwing a party, we did. Alfred Schild's widow, Winnie, often entertained the physics crowd. She had acquired a giant paella pan on their travels, and she would fill it with raw shrimp, olives, and sweet peppers, cover the whole thing in foil, and serve drinks while the marvelous dish cooked over an open campfire. Philip Bobbitt, too, was throwing open his new big house to wonderful dinner parties often for dozens of guests and often with local musical artists to entertain us.

When not partying, we might head for The Broken Spoke, a delightful joint. On the tables, there were babies in infant seats alongside mugs of beer. People in cowboy boots danced to country swing music. The vestiges of European forebears would show up in a spontaneous nightly group dance they called the "Schottish."

A special favorite of ours was County Line, a big but attractive restaurant. One branch was "the original;" another had water views. They could smoke a prime rib roast for eighteen hours and keep it rare.

In those days, all the barbecuing at the original County Line was done on site, and even the parking lot was redolent with the mouthwatering fragrance of smoked beef.

In 1981–82, with the Snells returned to Austin, Louise was living in Marge and Will Wilson's house. The Wilsons wanted to spend some time at their ranch in Cedar Park. Renting the Wilsons' house and getting to know the Wilsons was a wonderful introduction to Texas. They both have passed away now, alas, but they were charming, sophisticated, and kindly people. We had become fond of them and miss them. Will was a lawyer and a Republican politician, very much an Austin notable. Born in Dallas, Will Wilson was elected district attorney for Dallas County in 1946. He was elected to the Texas Supreme Court in 1950 and was Attorney General of Texas from 1957 to 1963. In 1969–71, he was Assistant Attorney General of the United States. He considered himself lucky that he had left the Nixon administration shortly before the Watergate burglary in June 1972.

The Wilsons' house was on Mt. Barker Street, high above Lake Austin. The air was full of bird song. Marge had placed bird-spotting binoculars at every window. In his study, Will had the best private Civil War library I have ever seen, with many regimental histories. On visits to Louise, I read Will's copy of Douglas Freeman's classic *Lee's Lieutenants*. I was glad to learn that Lee's most famous lieutenant, Thomas "Stonewall" Jackson, had been a physics professor at Virginia Military Institute before the war. But alas, Jackson was reputedly the worst teacher of physics in history. In his lectures, he would read to the class from the prescribed textbook, and if any student asked a question he would pause and start reading all over again from the beginning of the lesson.

The problem with all these wonderful arrangements and fun visits was that I was missing Louise anyway. Missing Louise, and, frankly, missing Austin, I found myself traveling most weekends. But I cannot say I enjoyed the time spent in making connections in Atlanta and Dallas–Fort Worth airports. So it was that, eventually, I sought and obtained a semester's visiting offer from the physics department at the University of Texas at Austin (UT), took the semester off from Harvard, and joined Louise in Austin. Freed from my onerous travel schedule and with Louise back in my life, I felt I could get some work done.

At UT, I offered the undergraduate course I had first given at Harvard in the spring of 1980, as part of Harvard's new core curriculum program. This course described the experiments that led to the discovery of the particles within the atom: the proton and neutron of which atomic nuclei are composed, and the electron. Instead of first setting out the principles of mechanics, electrodynamics, and thermodynamics that are needed to understand these experiments, I jumped right into the history, and brought in these principles and the maths needed to reason from them, only as they were needed. By making these physical principles part of a story, I hoped to give them an excitement that they would not normally have for a nonscientist. I thought that this might be a good new way of teaching physics to students whose main interests lay elsewhere. Everyone loves a story. Beyond that, I have come to believe that the best way of teaching mathematics is to teach it as part of courses on science, teaching mathematical methods only as they become needed. Louise's freshman astronomy course at Cornell had been taught in this way, and I may have been inspired by her experience.

At that time, the magazine *Scientific American* was setting up a new Scientific American Book Club. Its editor, Neil Patterson, invited me to write up my notes for this course as one of its series of books. Louise and I had taken to walking every evening in the Northwest Hills, and on one of these walks I told Louise that I needed a title for my book. She asked me what it was about. I told her it was about the discovery of subatomic particles. Without missing a beat, she said, "Well, call it 'The Discovery of Subatomic Particles.'" Louise was joking, referring to a favorite verse from the musical comedy *On the Twentieth Century*, a show we both loved. There, a pest approaches a producer and sings, "I have written a play,

Mr. Jaffe. It is about life in a large metropolitan hospital. I call it, 'Life in a Large Metropolitan Hospital' ...."

Joke that it was, in the end I could not think of a better title, and kept her joke title. As a book club selection, my book made lots of money, but when *Scientific American* needed to push newer books to members of their book club, they let mine languish. Fortunately, Cambridge University Press bought the rights, and in 2003 brought out a revised edition with fine new illustrations and with Louise's joke title.

I often benefited from Louise's inspirations in the matter of titles. For example, I was toying with "Explaining the World," when Louise came up with "To Explain the World."

During that spring semester in Austin, I took ice-skating lessons at a rink in the Northcross Mall. In Texas' usual hot weather, ice-skating was a fun thing to do. But I could not get the hang of it and fell, breaking an elbow. Soon afterwards, I traveled to Princeton to give that year's Henry Lecture. Friends at Princeton asked why my arm was in a sling. When I told them that I had broken my elbow while ice-skating in Texas, something about it seemed to them very funny.

My visit to UT, like all good things, had to come to an end. Back alone in our Cambridge house, and at Harvard in the fall of 1981, I received invitations to two meetings, very close in time and very distant in space. The first was to a Study Week on Cosmology and Fundamental Physics at the Vatican, from September 28 to October 1. The second was a meeting in honor of Geoff Chew at the Rad Lab in Berkeley, to take place on October 5. I arranged to attend both.

I thoroughly enjoyed the meeting at the Vatican. I had a pass that allowed me to get past the Swiss Guards and wander about the churches and gardens. I took my dinners in a nearby restaurant much patronized by clerics, which specialized in Sardinian food. There were interesting speakers, assembled by my friend, the Jesuit astronomer George Coyne. I spoke on elementary particle physics in the very early universe. There was a strangeness in trying to describe the first moments of the universe in scientific terms in this religious environment, but the good astronomers of the Vatican had found ways to be comfortable with modern astrophysics. And, after all, it was the physicists, far more than the clergy, that had to make themselves comfortable with the notion of a beginning: the Big Bang.

Participants in the Vatican meeting were invited to a papal audience at Castel Gandolfo after the meeting, but this would have prevented me from getting to Berkeley on time, so I passed it up. Chew had been very kind to me when I was at Berkeley, and I was not going to fail to give a talk in his honor. Missing Louise, I worked it out that I would fly from Rome to San Francisco on October 1 and 2, while she would fly to San Francisco from Austin on October 2, so that we could spend the weekend of October 3 and 4 together in San Francisco before I would have to drive over the Bay Bridge to Berkeley on the Monday.

It did not work. At the airport in Rome on October 1, I found that there was a strike of air traffic controllers, and my TWA flight to San Francisco via London and New York would not leave until the next day. I managed to make a telephone call to Louise canceling our tryst, and slept that night on a bench at the airport. At Heathrow, I found that there was another strike on, which made it necessary for the flight to New York to stop at Prestwick airport in Scotland to pick

FIGURE 13.1 The author with Geoffrey Chew in 1978

up supplies. I arrived at JFK airport in the middle of the night, scheduled to catch a flight to San Francisco early the next morning, on October 5, the day of the meeting at Berkeley. The hotels at the airport were all booked up, so I took a taxi to the Plaza Hotel, had two or three hours of sleep, and returned to JFK. With the help of time zones, I managed to get to the Rad Lab just about fifteen minutes before I was scheduled to talk.

Sleepy and unshaven, in my talk I said something like this: "These have been exciting times. Quantum field theory is riding very high, and one might be forgiven for a certain amount of complacency. But perhaps we will now see another swing away from quantum field theory. Perhaps that swing will be back in the direction of something like S-matrix theory." This was in part a way of honoring Geoff's work. I explained that effective field theory could be regarded as a fulfillment of the S-matrix theory program that Chew had pursued when I was at Berkeley in the early 1960s.

In the course of my numerous brief visits to Austin, and my semester visiting the UT physics department, I had come to like Austin and Texas' great university more and more. Both Louise and I found a geniality and friendliness in Austin that Boston, for all its receptiveness, could not equal.

Going to visit Austin on some February day, I would struggle through freezing, wet Boston. Surly airline people would hurl my bags on a runway while not responding to inquiries. I would get off the plane in Austin and step into a 70-degree day flooded with cheery yellow Texas sunshine. A smiling official would say, "How're you today?" and "Can I take that bag fer ya?" That evening, we would dine at County Line or go country dancing at The Broken Spoke. Life in Austin was more fun.

Members of the Harvard faculty could be friends and good collaborators with each other, but even so I had felt an intense competitiveness. Walking into the Harvard faculty club for lunch, I felt an almost electric tension in the dining room. At UT, things felt friendlier. We academics were working together, part of a great Texas tradition, and we weren't competing with anybody – except, perhaps, with other great research universities. We were ambitious for each other.

Austin had some very good physicists. In my fields, there were my former collaborator George Sudarshan and the prominent field theorists Bryce and Cecile de Witt. Cecile had been knighted by France for founding the Institute of Theoretical Physics at Saclay. There was the eminent John Wheeler, who had retired from Princeton, where I had taken his course on advanced quantum mechanics. (I had no interest in Wheeler's theories about black holes. How wrong I was.) The physics department also regularly received visits from the astrophysicist Dennis Sciama of King's College, London, and from Yuval Ne'eman of the University of Tel Aviv. Ne'eman had entered politics in Israel, and had been elected to the Israeli parliament, the Knesset. As a physicist, Ne'eman had made the proposal, at about the same time as Gell-Mann, of what Gell-Mann called the "eight-fold way," a symmetry that unites strongly interacting particles in families of eight or ten members. There were other well-known figures at UT, whom I will be discussing later.

I was also impressed by the mixing of the university community with the rest of Austin. At parties, I met not only professors but also politicians, business leaders, judges, journalists, writers, and ranchers. This was a change from Cambridge, where almost all our friends were connected to the academic community. Of course, Cambridge has quite an academic community, perhaps the most outstanding in the world. But even so, in Cambridge, Louise and I felt cut off from the nonacademic life around us. After more than fifteen years in Cambridge, we were just beginning, through the American Academy and the Saturday Club, to have a few friends outside the academic world, and appreciated it. But the experience was much more readily had in Austin.

Also, at that time, Austin was much smaller than it is today. When we joined the audience at one of Austin's many little theaters or boarded a plane at the Austin airport, we would almost always see friends. Speaking of the little theaters here, I should add that now, having spent forty years in Austin, and having enjoyed its many little theaters, so much more numerous and lively than those in Cambridge and Boston combined, we have grown up and grown old with its talented corps of local actors. Over the years, they have moved us to tears or laughter and shaped an important part of our lives.

But apart from the attractions of Austin, I was miserable being separated from Louise. Returning alone to our empty house every day was a dreadfully depressing ordeal for me, and waking up to an empty house was almost as bad. We were not only a couple; we were comrades. And we had had the shared experiences of our thirty years together. It is hard to describe the emptiness I felt. I had never imagined that I could give up what I had at Harvard, but I found myself thinking of it seriously now – giving up the Higgins Professorship, the office with the fireplace, Faculty Council meetings in the Bulfinch room, the highest salary on the Harvard pay scale, all of it. And then there were the many friends in Cambridge, and the interesting people I was meeting at my clubs. And none of it seemed to matter.

So it was that I found myself negotiating with Tom Griffy, the chairman of the physics department, for a possible move there from Harvard. This was not easy for UT. I had a very good arrangement for salary at Harvard, the equivalent of a University Professor's salary, which, for bureaucratic reasons, was difficult to match at Texas. An effort would have to be made to establish a position for me which I could accept, something roughly equivalent to what Harvard was providing. So Griffy thought it would be helpful to put the problem to Peter Flawn, the then President of the UT. I later learned, in a similar effort to create a home for Roy Schwitters, my dear friend, and then Director of the proposed Super Collider, that, on occasions of this kind, people at the University often call on the resources of the whole community, and Austinites of means or influence do step up to the plate. This dynamic community feeling remains a great feature of professional life in Austin.

While we were working out the arrangements, my impression of this community feeling was strengthened and extended from Austin to the state, in effect, when a physics professor, Bill Drummond, took me to the Capitol to meet the master of the Texas Senate, Lieutenant Governor Bill Hobby. We had a long talk about Texas, during which Pete Laney, the Speaker of the Texas House, dropped in for a meet and greet. What really impressed me was that a physics professor had been able to set it up.

Bill Hobby became a friend of ours, and was a friend to science in Texas. He was a major donor to the Hobby–Eberly Telescope at McDonald Observatory, one of the world's largest. Astronomers are now using it to study the motions and distances of thousands of galaxies as a means of learning about a mysterious dark energy that is causing the expansion of the universe to accelerate. (More about this later.) In 1997, I would be a proud speaker at the dedication of the Hobby–Eberly Telescope.

In the end, the University and I came to a happy arrangement. The Welch Foundation endowed a chair for me, the Josey Regental Chair in Science, in honor of Jack Josey, a longtime civic leader. It was denominated a chair in science because the Welch Foundation generally supports work in chemistry rather than physics, and "science" nicely spanned both disciplines. Unlike the Higgins Professorship at Harvard, this chair was not purely honorific. Wonderfully, it came with research funds that I could use to support the work of a Theory Group that I would head. There was a further attractive feature of the package. In addition to joining the physics department, I would be a member of the very good Texas astronomy department. The University's astronomers run the McDonald Observatory in West Texas, at that time under the directorship of Harlan Smith.

I accepted this offer with a good feeling of welcome. And this warm welcome would enable me to live my life with Louise. Although I would officially be on leave from Harvard through the 1982–83 academic year, I began my Texas professorship in Austin in 1982. For a while, a rumor went around Austin that I had been afforded a larger salary than the football coach. That of course was absurd. No professor at the UT would ever have a higher salary than the football coach. But, as I explained to friends, I would have tenure, and coaches do not. My job would not be contingent on winning football games.

With this settled, we had to find a place of our own to live. Marge and Will were willing to let us have their place as long as our house-hunt took, but after a very brief search we bought a wonderful woodsy house on a nature preserve. The house had a pool. Louise bought it on the spot while I stood silently by. She had seen at once that it was just the ticket for us.

We held on to our beloved house in Cambridge, and used it in summers and over the holidays, and for visits to Elizabeth during the academic year. We still have that house, although we can no longer imagine compressing our lives down to it.

In 1982–83, I was still officially a Harvard faculty member on leave, so during that year I would occasionally attend the Harvard physics department faculty meetings. At the last meeting that I attended, I urged the department to try to hire Frank Wilczek as my replacement. Though we never collaborated, our work in physics seemed to follow parallel grooves. We had come up simultaneously with the proposal of the axion particle, and Frank with Tony Zee, at the same time as me, had cataloged terms to be added to the equations of the Standard Model that would violate baryon conservation. I think the offer was made, but unfortunately for Harvard, Frank decided instead to go elsewhere. These days, I see his picture now and then decorating a fine column he occasionally writes for the *Wall Street Journal*. It was a great mischance for Harvard.

I handed in my formal resignation from Harvard early in 1983. The Higgins Professorship, along with the paneled office and marble fireplace, fell to Shelly Glashow.

The years at Harvard had been good for me, but there is life after Harvard – I have found my time in Austin even more interesting.

To begin with, I was early asked to join the Headliners Club in Austin, which has contributed much since then to the fun of being in Austin. Like the Petroleum Club of Houston and other private down-town clubs in Texas cities, it occupies the top floors of a tall bank building. From its large windows, there are views of the Capitol and university tower to the north, Town Lake to the south, and to the west the beginnings of the Texas hill country. The Club hosts dinner meetings and lunch meetings for groups of all sizes, from half a dozen to several hundred, and has several wonderful bars. One gets to know and appreciate the skilled bartenders and attentive managers and wait people. The Headliners Club was founded in 1954 as a press club, but then expanded to include Texans who might be the subjects of news stories as well as those who report about them. I was on the Board of Governors of this foundation from 1994 to 2017, and in this way had the pleasure of getting to know quite a few interesting Texas journalists and politicians.

When I first moved to Austin, the Club was full of people who had served in the Johnson administration, and we came to know and like so many of them. I will write about them in a future chapter. Here I will simply say they were very welcoming to us. Alas, we have outlived most of those good people.

In 1982, I became worried about problems faced by Israel. There was widespread press condemnation of Israel for its attack on the Palestine Liberation Organization (PLO) in Lebanon in June 1982. It seemed to us that this was unjust. The PLO had been attacking Israel from Lebanon for years, just as Hezbollah has been doing latterly.

After all, the PLO had been founded in 1964 when no part of the West Bank or Gaza was in Israeli hands, so for the PLO, "liberation" could only mean the elimination of the State of Israel, an aim still acknowledged. Israel's attack on the PLO in Lebanon was clearly in defense of its civilian population, and, indeed, its existence. There was nothing about it that should have invited condemnation. This condemnation was an example of world intolerance of the little Jewish democracy, and disregard of its security needs. It is to my eyes a continuation of centuries of world enmity to the Jewish people. One might have hoped that the bloodthirstiness of the Holocaust and of PLO hijacking and terrorism would have produced some sympathy for the plucky little Jewish nation. But even Jewish Americans obtusely sympathize with the enemies of Israel, and it has become fashionable on campuses and in cities all over to hate Israel and love its medieval attackers. The perverse narrative of noble "resistance" is believed, and beleaguered Jewry is still hated. It never ends.

Yet Israel is the only democracy in its part of the world, an example and a light. And, with the surprising success today of the Abraham Accords, Israel is making some friends among its Arab neighbors, and all over the world. Still vilified, and still positive and plucky. As I write this, Israel is in existential danger of annihilation at the hands of Iran, which is clearly approaching breakout as a nuclear power. Iran's leaders say of Israel that it is a one bomb country. One bomb, that is, is enough to destroy all of it.

I was casting about for something we could do that might – in a small way – be helpful to Israel, and would show our support. My first idea was to start a summer school of theoretical physics to meet in Israel's capital, Jerusalem. This would be a summer-long festival conference of talks by experts in various fields of physics. But, mindful of summer heat in the Middle East, I proposed instead a winter school. It would meet briefly when most universities were closed, over the winter holidays.

I had a great deal of help from my friend Yuval Ne'eman, who was then Israeli Minister of Science and a frequent visitor to the UT. He and his beautiful wife D'vorah were splendid hosts to the capital city, its famous churches, and the Knesset. I also was much helped by Tsvi Piran of the Hebrew University in Jerusalem, to whom I was introduced by Ne'eman. Tsvi would eventually put the annual conference under the auspices of Hebrew University. We planned for the first Jerusalem Winter School to run at the Givat Ram campus of the Hebrew University in Jerusalem from December 28, 1983 to January 6, 1984 (as I told my Israeli friends, from Christmas to Epiphany). There was to be a rotating directorship, but today, in its 39th year, the Jerusalem Winter School is under the nominal permanent directorship of David Gross, a great theoretician and Nobel laureate in physics, working at Santa Barbara.

The lectures at the first convening of the Jerusalem Winter School were on the intersection between elementary particle physics and cosmology, and were given by leaders in the field: Alan Guth of MIT, Jim Gunn of Princeton, Stephen Hawking of Cambridge, and Mike Turner of Chicago, plus me. We had invited Hawking chiefly as a courtesy, aware that his physical condition would make the trip impossible, but were delighted when he came to lecture, accompanied by helpers who when necessary carried him and his wheelchair from place to place.

With Louise, I would journey to Jerusalem every few years, always enjoying gracious shepherding and hosting by the Ne'emans, Yuval and D'vorah. Yuval was then a stalwart conservative member of the Knesset, and they would often visit us in Austin. Alas, like so many of our much-loved friends, here and abroad, the Ne'emans have passed away.

The Jerusalem Winter School has been a great success. I continued as director through the first seven winter schools, running up to the 1989–90 school, and lectured at several of them. We stayed in a small apartment at Mishkenot Sha'ananim, the stone-walled garrison guest house of Jerusalem that in 1860 had been built as an almshouse by Sir Moses Montefiore. It was built to withstand Arab attacks. It had only a few guests at a time, often interesting ones. We were there introduced to the pianist Menachem Pressler, who had once considered moving to Texas, and to the anthropologist Claude Levi-Strauss. I asked Pressler if the acoustics in the various auditoriums at UT were the problem, but, amusingly, he said, No, the acoustics were fine, he simply couldn't stand the Dean.

In the 1980s, we could enjoy ourselves wandering around the crowded narrow streets of the Old City in Jerusalem – that part of the city encircled by ancient stone walls, with its famous gates. We had not been able to do so during our early visit to Israel in 1961. Until 1967, the Old City had been controlled by Jordan. It had been cleared of Jews by the Jordanians and was Judenrein, as the Nazis would say, and inaccessible to anyone arriving from Israel except for one day each year, at Easter, when Christians might be permitted grudging entry. Jewish habitation was forbidden in the Old City until 1967, as in Jordan itself to the present day. The world seems disregardful of the Nazi arrangements in most Arab countries. I especially remember in the 1980s, under Israeli security at last, being allowed to visit two beautiful buildings: the Al Aqsa mosque on the Temple Mount, and the crusader church of St. Anne. We had the pleasure one winter of having the company of friends from Cambridge on these walks: Dan and Pearl Bell, and their son David. We visited the Knesset with Yuval, where he was a popular Likud member, and he and D'vorah took us to a Greek Orthodox Church to see the lighting of the candles and hear special Greek Orthodox carols. Interestingly, when Louise expressed personal dislike of the peculiar garb of the Orthodox Jews in Jerusalem, D'vorah said, with characteristic wisdom, "But of course those are the very ones who need safe harbor."

One winter, we took a flight from the small Jerusalem airport to Eilat, on the Gulf of Aqaba, Israel's only port east of Suez. We stayed in the suburb of Taba. At that time, Taba was a chic beach resort, and a scene of fashion. It would later become the "pencil dot on a map" that had to be turned over to Egypt. In an underwater observatory, built into a coral reef, we had a view of spectacularly colored tropical fish.

On another memorable occasion in Israel, we were taken to the freshwater springs at Ein Gedi on the Dead Sea shore. Looking east across the water, we saw low mountains colored pink by the setting sun. When I asked what they were, we were told that these were the Mountains of Moab. Biblical antiquity pervades the land of Israel.

We also drove by ourselves around Israel and the West Bank, and had a favorite place for lunch under orange trees in Jericho. We visited the Dead Sea and I floated atop its buoyant salt water.

In just one respect, the Winter School was a disappointment. We had hoped that it would bring in young physicists and physics students from neighboring Arab states. That did not happen. I tried several times to get Abdus Salam to lecture at the Winter School, but he told me that it would be impossible; if he visited Israel, he would lose the support of Muslim countries for his own institute in Trieste.

We gave the Jerusalem Winter School a good start. The school has continued to thrive to the present. Hebrew University has built on that foundation to offer Winter Schools in a variety of other subjects. The founding of the Jerusalem Winter School in Theoretical Physics is one of the things of which I am most proud.

In the early 1980s, I worked on the implications of some attractive speculative ideas. One was supersymmetry. The symmetries of the Standard Model only join particles of the same spin into families: spin ½ quarks with other quarks, spin one photons with spin one W and Z particles, and so on. A supersymmetry is a symmetry that unites families of particles of different spin. For most theorists, awareness of the possibility of supersymmetry began with a quantum field theory presented in 1974 by Julius Wess and Bruno Zumino (though unknown to Wess and Zumino, a model with broken supersymmetry had already been published by D. V. Volkov and V. P. Akulov).

These papers just showed what is possible. The theories they presented were not intended to be realistic. Indeed, there is no sign of supersymmetry in the menu of known particles. It has been joked that supersymmetry is a symmetry that unites every known particle with an unknown particle. In proposed supersymmetric theories, spin one photons, W and Z particles, and gluons are put by supersymmetry into families with hypothetical spin ½ photinos, winos, zinos, and gluinos. (The "ino" is taken from a particle of spin ½, the neutrino.) Likewise, spin ½ quarks and leptons are put by supersymmetry into families with hypothetical spin zero squarks and sleptons. (The "s" is used here because the fields of particles of spin zero are of the type known as scalar.) Murray Gell-Mann once referred to this as a "slanguage"; I have instead called it a "languino." Because there are evidently no squarks, sleptons, photinos, winos, or zinos with the same mass as quarks, leptons, photons, or W or Z particles, it is necessary to suppose that supersymmetry is another broken symmetry, like the symmetry uniting weak and electromagnetic interactions. It is hoped that the new particles it requires are just too heavy to have been created at existing accelerators, and will be found when experiments can be pushed to higher energy. So far, this has not happened.

I wrote some papers on the crucial question of how supersymmetry is broken, and on its implications for cosmology. But this has remained only an exploration of possibilities. Supersymmetry continues to be attractive to theorists, in part because a symmetry that unites particles of different spin is nearly impossible once one starts in this direction and assumes something about the content of the theory, one has hardly any freedom in constructing a theory. So at least one winds up making definite proposals.

The other speculative idea that I worked on in the early 1980s was much older than supersymmetry. In 1921, Theodor Kaluza in Germany attempted a unification of gravitation and electromagnetism by rewriting Einstein's general theory of relativity as a theory in five space-time dimensions. The gravitational field in Einstein's theory is described by a set of quantities $g_{\mu\nu}$ depending on position in space and time, whose labels μ and ν both run over the three directions of space and the one direction of time. In Kaluza's theory, the indices can also take the value 5, labeling the direction along the new fifth dimension. The quantities $g_{5\nu}$ in his theory are the potentials whose rates of change are the electric and magnetic fields, while the quantity g_{55} is the field of a new sort of matter.

For Kaluza, this was purely formal, a matter of juggling indices. A few years later, it was given a more physical interpretation by the Swedish theorist Oskar Klein. He considered a five-dimensional space-time in which one dimension is tightly rolled up, so that inhabitants of the space-time who do not probe very short distances experience it as four-dimensional, just as a being, living on the two-dimensional surface of a drinking straw, might think its space is one-dimensional if it does not probe distances as small as the straw's diameter. Einstein was taken with these ideas, and used them in his unsuccessful search for a theory that would unify general relativity with Maxwell's theory of electromagnetism.

We now know that there are more forces in nature than just gravitation and electromagnetism. To recover the strong and weak interactions of the Standard Model in a theory of the Kaluza–Klein type, it is necessary to introduce a space-time of more than five dimensions. Since the 1970s, this idea has been pursued particularly in the context of string theories, which find their natural formulation in ten space-time dimensions, with six dimensions tightly rolled up. In the early 1980s, I worked out how to calculate couplings like the 1/137 of electrodynamics from the circumferences of the rolled-up surface, and spoke about this at a Solvay Conference, at a meeting at the Royal Society of London, and also at a meeting at Shelter Island.

I also considered theories of gravitation in higher dimensions that are even more speculative. In these theories, there is an equivalence of gravitation and inertia, as in general relativity, but the symmetry that applies in inertial frames of reference – such as freely falling elevators – is more general than just higher-dimensional generalizations of the symmetry underlying special relativity. I talked about this in the second Jerusalem Winter School.

My interest in physical theories in higher dimensions has continued. At the time of writing, my latest research article is a 2020 paper describing the field theory of massless particles in space-times of any dimensionality. But nothing in the way of a realistic theory that could be tested experimentally has come out of any of this.

Some physicists disparage this sort of speculation. They remind us solemnly that physics is an experimental science. Yes, it is, and physics is never more progressive than when theories are being suggested and tested by experiment, as in the 1970s. But as I said in the opening of my Nobel lecture, in the dark times between experimental breakthroughs, there continues a steady evolution of theoretical ideas. It is these ideas that may make possible the next step. A good example is provided by theories of the Yang–Mills type, in which photon-like particles of spin one interact with charges that they themselves carry. When first proposed in 1954, these theories seemed to have no relevance to experimental reality. But after the proposals of spontaneous symmetry breaking and infrared slavery, they became one of the foundations of the experimentally successful Standard Model. Speculations about supersymmetry or higher dimensions may turn out to have been a waste of time, as was once thought of Yang–Mills theories, but it is an essential part of the craft of theoretical physics to be willing to risk wasting time.

In the spring of 1983, I resigned formally from Harvard. We traveled to Oxford, where I gave the annual Cherwell–Simon Lecture. (Lord Cherwell was an Oxford physics professor who served as Churchill's longtime science advisor. Francis Simon was a physical chemist and Cherwell's successor at Oxford.) We stayed at the Randolph, of course.

There was a dinner for us at All Souls. It should have been pleasant, but there was a snobbish person who made it clear that he did not welcome Jewish Americans, and greeted us by addressing another member of the house and saying, in our hearing, "Get them out of here. Show them the garden." At dinner, he said to Louise, "You see, we don't like you." An awful silence fell at that table. Louise said, slowly, "I am very sorry to hear that. Up to now, everybody has been so kind." As if that were not enough to put him in his place, she contributed to his further disintegration on taking her leave. We had learned that he was a musicologist, and in exiting she said, "Our neighbor is Leon Kirchner, the composer. We will tell him of you, he will be so interested." The oaf reached out, stuttering, and tried to detain us, but we were out the door.

I had thought that one benefit of the move from Cambridge to Austin is that I would stop being tempted to spend time on interesting committees. Alas, the temptation and the calls of duty were too strong. Starting in 1982, I served on the Council of the American Academy of Arts and Sciences, the Council of Scholars of the Library of Congress, the Board of Governors of Tel Aviv University, and many conference organizing committees, award committees, and editorial boards. Keeping up my involvement with arms control, I joined the Committee on International Security and Arms Control of the

National Research Council, chaired by Murph Goldberger. I remember well Murph's seeing me there for the first time since I had gone to Texas. He greeted me with a hearty "Howdy!"

One conference that I attended and helped to organize was inspired by history. I refer to the historic June 1947, Shelter Island Conference. At the start, the Rockefeller Institute in New York City had convened a meeting at the Ram's Head Inn on Shelter Island, at the eastern tip of Long Island, to bring together physicists who, after the war, were ready to resume work on the fundamental problems of physics. Present were leading physicists of the 1930s, including Hans Bethe, Gregory Breit, Hendrik Kramers, I. I. Rabi, Viki Weisskopf, John Wheeler, and Robert Oppenheimer. Oppenheimer chaired.

The electron theory of Dirac had predicted that a pair of the excited states of the hydrogen atom would have precisely the same energy. It had been guessed that the energies of these states would be slightly split by so-called radiative corrections, due to processes in which the electron, while in orbit around the hydrogen nucleus, would emit and reabsorb photons. But calculations of this splitting found that it would be infinite, not a plausible result. In consequence, many physicists concluded that, for some fundamental reason that was not yet understood, the splitting of the energies of these states due to radiative corrections must actually vanish. No one had as yet detected any splitting of the energies. It was therefore exciting when there, at Shelter Island, Willis Lamb announced that, using radar techniques developed in the war, he was able to detect an energy difference between these two states of hydrogen. He described the value of this energy difference by reporting that, for radiation to produce transitions between the states, its frequency must be about 1,000 megacycles per second – a tiny frequency by the standards of atomic physics.

When I was a graduate student in Copenhagen, this surprise was summarized in a slogan, "Just because something is infinite does not mean that it is zero!"

Lamb's measurement was a challenge flung to theorists, to go back to the calculation of what was now called the Lamb shift. After the meeting, on a train ride to a consulting commitment at Schenectady, Hans Bethe did a rough nonrelativistic calculation. He found a near cancelation of the radiative corrections, in which a single photon is emitted and absorbed, and those in which the photon is accompanied by an electron–antielectron pair. The result was still infinite, but when it was made finite by imposing a limit on the energies of the particles emitted and absorbed, it turned out that the energy-splitting now depended only very weakly (logarithmically) on the limit imposed. Bethe more or less arbitrarily took the limiting energy to be half a million electron volts, the energy contained in the mass of an electron, and found an energy-splitting corresponding to a photon frequency of about 1,040 megacycles per second. A year later, by using the idea that infinities could be absorbed into a redefinition or "renormalization" of the electron mass and other constants, an accurate calculation with no arbitrary energy limit was done by Viki Weisskopf with J. B. French, and independently by Lamb himself with Norman Kroll. They found a result of 1,052.19 megacycles per second, in good agreement with more precise measurements.

The near cancelation of radiative corrections found by Bethe is actually a consequence of the principles of special relativity obeyed by quantum electrodynamics, but this was thoroughly obscured by the old-fashioned calculational techniques used by Bethe and then by French and Weisskopf and by Kroll and Lamb. To deal effectively with infinities in other calculations, it was necessary to invent new techniques that would exhibit a manifest consistency with relativity at every state. This was accomplished in different respective ways by two participants at Shelter Island, Richard Feynman and Julian Schwinger. They reported on their work at a follow-up conference in the Pocanos a year later.

Revering this history, Nicola (Nick) Khuri of Rockefeller University, together with Roman Jackiw, Edward Witten, and me, organized a conference to meet in June 1983, in the same room of the Ram's Head Inn on Shelter Island. Among the participants were some of the heroes of 1947: Bethe, Feynman, Lamb, Marshak, Rabi, and Weisskopf. In the spirit of the original Shelter Island Conference, there were present also younger physicists who had been playing leading roles in theoretical physics in recent years, including Stephen Adler, Michael Duff, Murray Gell-Mann, David Gross, Alan Guth, Stephen Hawking, Toichhrio Kinoshita, T. D. Lee, Andrei Linde, Yoichiro Nambu, K. Nishijima, John Schwarz, Peter West, and Bruno Zumino, along with the organizers.

This was a conference on "Quantum Field Theory and the Fundamental Problems of Physics." As had been intended from the beginning, the participants at this Shelter Island II now also included a later generation of elders, including Richard Feynman, Willis Lamb, Robert Marshak, and Julian Schwinger. It was their presence that turned out to be most important historically.

Marshak (who would be my host at Rochester two years later, in 1985) dispelled a cloud of confusion that had grown up regarding

FIGURE 13.2 The Shelter Island Conference, 1983

cosmic ray phenomena. He argued that there are two particles of similar mass that had been previously confused with each other. There is the pion, a strongly interacting particle produced copiously in the collisions of cosmic ray particles with atoms in the air. There is also a slightly lighter particle that does not participate in strong interactions, the muon, into which the pion decays with the emission of a neutrino.

No breakthrough occurred at Shelter Island in 1983, nothing as dramatic as the announcement of the Lamb shift in 1947. We had to pay the price of our success. There was no exciting challenge to the Standard Model. My most vivid memory is from the trip back from the meeting. After our visit to Oxford in May 1983, Louise and I had flown to Boston and stayed at our house in Cambridge. In June, we drove from there to a port on the Connecticut shore of Long Island Sound and took a ferry to Long Island. After the meeting, Feynman was with Louise and me on the ferry back to Connecticut, together with other physicists, including (as far as I remember) Guth and Jackiw. We had an enthusiastic loud conversation about physics, using the jargon common among physicists. A fellow passenger on the ferry became angry, and said something along the lines of "You guys think you're so smart!" We continued our conversation, a little more quietly, but Feynman went over to our angry fellow passenger and had a long talk with him, which I could not hear. At the time, I thought that Feynman was showing that he was a regular guy, not a pointy-headed loud intellectual like me and the other physicists on board. I already felt a dislike for Feynman, going back to his reception of a talk I gave at Cal Tech when I was on the faculty at Berkeley. My view of Feynman softened later when I read in the Feynman biography by James Gleick how Feynman supported his wife Arlene as she lay dying during the war, and of his pain at her death. I began to see him as human, not just a brilliant brat. Now I wonder if perhaps on that ferry he was trying to soothe the feelings of a man that had been hurt by his hearing a conversation that he could not understand.

In the early 1980s, a great financial crash caused Austin to lose many of its local banks. Our own bailout was arranged by an official sent by the Federal Deposit Insurance Corporation (FDIC) to Austin. Now Austin banks are mostly branches of big national banks. True children of the Depression of the 1930s, Louise and I both depend on simple bank accounts for savings, preferring security to investment income.

In 1984, Louise was made a member of the American Law Institute (ALI), the closest thing in law to the National Academy of Sciences. The ALI prepares restatements of the law in various areas. She has been active in advisors' debates at the ALI headquarters in Philadelphia, which has brought us many times to stay in hotels at Rittenhouse Square, one of America's best urban spaces. In 1985, she was named Raybourn Thompson Professor of Law at Texas. Eventually, she would hold the William B. Bates Chair in the Administration of Justice, the chair formerly held by Charles Alan Wright.

After my arrival at the physics department in Austin, I had the responsibility of recruiting theoretical physicists to our new Theory Group. My first selection was Willy Fischler, who had been strongly recommended by Lenny Susskind. It was an excellent choice – Willy has made important contributions to a broad range of topics, from the axion, mentioned in Chapter 12, to quantum gravity. With Willy's advice in 1984, we recruited Joe Polchinski, another very successful choice. Joe made a large splash in 1989 with a contribution to string theory, more about which later. Like Willy, he was exceptionally versatile. For example, he applied effective field theory to justify approximations made in the modern theory of superconductivity. Losing Joe Polchinski in 1992 to the Kavli Theoretical Physics Institute at Santa Barbara was my biggest failure as leader of the Theory Group. The Theory Group has continued a slow expansion, recruiting other bright theorists: including Vadim Kaplunovsky, Can Kilic, Sonia Paban, Jacques Distler, and Katie Freese.

In 1984, I was elected to the Philosophical Society of Texas. This society was founded in 1837, the first year of the Republic of

Texas, by Sam Houston and his friends. Its name was inspired by the name of America's first learned society, the American Philosophical Society. The Texas society became inactive after a few decades, but was revived in 1937. Since then, it has held meetings once a year, moving from place to place in Texas, where members hear talks on a topic chosen by that year's president – and use the occasion to reconnect with friends. Its membership includes academics, politicians, authors, artists, lawyers, business people, and even one or two philosophers. The one limitation is that members must be residents of territory that was once part of the Republic of Texas, which includes not only the state of Texas but a strip of land extending up to Aspen, Colorado.

Louise later became an elected member of the Philosophical Society of Texas as well, nominated by Elspeth Rostow. Participation in the meetings of the Philosophical Society has taken Louise and me to many places in Texas that we would likely not otherwise have had occasion to visit, including Abilene, Corpus Christi, Fort Worth, Kerrville, Laredo, and Waco, as well as cities like Dallas and Houston and San Antonio, that we had visited often.

Louise headed a panel on the arts at the Philosophical Society meeting in San Antonio in 1989, and in Waco in 2017 gave a talk on "The Economic Rights of Individuals." This last won the special thanks of the club's President; Louise had enlivened a perhaps overly earnest morning with hilarious commentary and a colorful slideshow.

I served as President of the Society one year, organizing a panel of speakers. I had the good fortune of securing Stephen J. Gould in one of his last public appearances.

As far as I know, there is no Philosophical Society of California, or of Florida, or of any other state. This is one example of an attachment of Texans to their state that I have not seen elsewhere in America. If you ask an American where he comes from, he may say San Francisco or Boston, but is much less likely to say California or Massachusetts, while someone from Houston or Dallas is likely to reply, "Texas." There is a lot about Texas state politics that I would wish to change, but I have enjoyed a feeling here of living in a new country, without in the least having left America.

The W and Z particles of the electroweak theory were discovered at Conseil Européen pour la Recherche Nucléaire (CERN) in 1983, with masses respectively about eighty-six proton masses and ninety-seven proton masses, just as predicted by the theory. With unusual speed, the Nobel Prize was awarded to Carlo Rubbia, who led the team making the discovery, and to Simon van der Meer, who had made this discovery possible by inventing a brilliant way of focusing beams of particles in accelerators.

Since that year's laureates had confirmed my predictions, the Nobel Foundation invited me back to Stockholm to join the festivities that December. It was so nostalgic for Louise and me. It was delightful to return to beautiful snowy scenes and Swedish Christmas carols that we had come to love.

In 1985, I joined the Town and Gown Club in Austin. It was a small group of clubbable people who enjoyed getting together for a seated dinner and a talk, usually by a member. At that time, it was all male, but I was determined when I accepted membership to "liberate" the Club and get interesting women into it. As its name indicates, this Club brought together men from the university and the rest of Austin, just the sort of thing I liked about life in Austin. It meets at the Headliners Club for dinner and a talk by a member or visitor once a month, and in May for an outdoor barbecue. In my first year, the barbecue was at Will Wilson's ranch.

The Club's exclusion of women also bothered another club member, Arnold Rosenfeld, then editor of our local newspaper, the *Austin American-Statesman*. Rosenfeld and I quit the Town and Gown Club, and resolved to found a similar club that would admit women members on the same basis as men. Early in 1986, we convened a dinner meeting of six couples, for the purpose of laying plans. Those present besides Ruth and Arnold Rosenfeld and Louise and me were Nancy and Admiral Bobby Rae Inman, Marian and Hans Mark, Elspeth and Walt Rostow, and Judges Mary Pearl and Jerre Williams.

We decided to hold seated dinner meetings on the third Tuesday of every month during the academic year, a day chosen because Jerre Williams would then usually be in Austin despite his service on the United States Court of Appeals for the Fifth Circuit, sitting in New Orleans. With the Saturday Club of Boston somewhat in mind, we adopted the name the "Tuesday Club." Judge Mary Pearl Williams agreed to write the bylaws. The bylaws ensured the perpetuity of the Club by its arrangements for past presidents' meetings that would select the following year's president, usually alternating between men and women. Louise's important contribution was the provision for annual turnover in the Presidency. Groups we had joined often faded away when a leader of many years died or retired or moved to another city. It was agreed to offer membership to couples who, like the ones at the founding dinner, would both make interesting contributions to our dinner meetings, and to try for a balance between downtown Austinites and university people in the membership. Before the pandemic struck, the Tuesday Club, a great success, had over 200 members and a branch in Atlanta. We would meet at Headliners' beautiful lounge, obtain drinks from its skillful, friendly bar, and go to more or less randomly assigned tables, where one would find friends old and new.

Meanwhile, perhaps shamed by the success of the Tuesday Club, Town and Gown at last decided to become open to women members. About fifteen years ago, Louise was invited to join Town and Gown. She is now the valued member, and I the spouse. She has given three invited after-dinner talks at Town and Gown, on foreign atrocities, Gilbert and Sullivan, and the economic rights of individuals, each time enlivening the evening with one of her amusing slide shows.

We both became habitual attendees of Roger Louis' British Studies program at the Harry Ransom Center, enjoying Friday afternoon sherry and talks by foreign visitors and some locals. Louise gave several talks to that group as well. When Roger had some distinguished English visitor, and questions from the audience might flag,

Roger came to count on Louise as a dependable resort – she always had something interesting and pertinent to say, to which the visitor would enjoy responding. I gave a talk about Newton.

In 1985, I received an honorary doctoral degree at the commencement of Washington College, a small liberal arts college on the eastern shore of Maryland. It was an especially gratifying occasion for me because the degree was a DLitt instead of a PhD or ScD. In the early 1980s, I had received honorary degrees from Clark University, Dartmouth University, and the Weizmann Institute, but this was to be the first time that I would be the commencement speaker as well. I took the occasion to take some swipes at the "Star Wars" antimissile program of the Reagan administration, but my main topic was science education. I ended with a comment: "I know of no better way of teaching science to undergraduates than through its history. Science is, after all, a part of the history of humanity, and not, I think, its least interesting part."

But the most interesting part of this visit was not my talk, but the dinner Louise and I had the previous evening with the college president Douglass Cater and the other honorand, the violinist Isaac Stern, and his wife Vera. We were put up at the president's house, and we stayed up late talking with the Sterns about music. At one point, when I mentioned a piece of violin music, Stern said that he did not know it. At first, I could not believe it, and then I realized that he had meant that he did not know it well enough by heart to play it without a score. I asked Stern how it is that music that has no programmatic content, like Beethoven string quartets or the Shubert string quintet, could have such a powerful emotional effect on us. He thought that it had something to do with the physiological processes implanted as the fetus grows near its mother's heart in the womb.

Stern had to get somewhere to give a performance the next day, so after the commencement, Isaac and Vera, and Louise and I were put in a small plane that flew from a grass strip near the campus to Baltimore–Washington airport. I thought it a risky flight, but Stern was very good company. Louise asked Stern which of the great violin concertos – Beethoven's, Brahms', and Mendelsohn's – was his favorite. He thought about this and confessed that he loved them all and never tired of them.

At the end of June 1985, we went to the Canary Islands, to participate in the inauguration of the observatory at Roque de los Muchachos, on the rim of a volcanic caldera on the island of La Palma. For optical and infrared astronomy, this site is second only to Mauna Kea in the Northern Hemisphere. For that reason, many nations have placed large telescopes at this site, starting with the Isaac Newton Telescope, moved there from Britain. We stayed at a hotel on Tenerife, the administrative center of the Canaries. On Tenerife, I gave a talk on the relations between particle physics and astrophysics at the University of La Laguna. Then, before dawn on June 29, we were awakened, and, with other visitors, were flown on a Spanish military aircraft to La Palma, where we boarded a bus that drove us up a winding mountain road to the observatory, almost 8,000 feet above sea level. There we were asked to wait on a lawn covered with folding chairs, under the unshaded sun, for royal persons from Britain, Denmark, Spain, and Sweden to arrive from Tenerife. These distinguished royals had been invited to demonstrate sovereign commitment to the new telescope. Looking down from the rim of the caldera, we could see the volcanic bowl filled with sea fog, while at our altitude we were in bright June sun under cloudless skies. Much too much sun, in fact. Louise and other visitors fled indoors, while I tried to find shade outside and a party of Swedish astronomers on lawn chairs enjoyed the rare treat of enough sunshine. After a few hours' wait, tired, disheveled, hot, and thirsty, we heard helicopters arriving. They brought the royal persons, all of them cool and immaculate – the women in enormous hats, as in Cecil Beaton's design for Ascot opening day in *My Fair Lady*. After their brief flight from Tenerife (no bus toiling uphill for them), they were led to a special facility for refreshments. From this experience, I learned the lesson, never to attend any ceremony attended by royal persons.

Around that time, I began to be seriously interested in string theory, a major preoccupation of members of the Theory Group I had put together at the UT. In string theories, the fundamental ingredients of physical theory are not particles or fields; they are strings, one-dimensional discontinuities in space, and their many modes of vibration recognized by us as different species of elementary particles. String theory had its start in 1968, when Gabriel Veneziano, a young theorist at CERN, working on strong interactions, suggested a simple analytic formula for a meson scattering process that satisfied many of the requirements of S-matrix theory. Several theorists, including Mandelstam and Nambu, soon recognized that the spectrum of particle masses in Veneziano's model corresponded to the modes of vibration of a string. These particles were all particles of integer spin, like mesons. It was in the effort to bring particles like quarks, of half-integer spin, into the theory that supersymmetry first appeared – though not yet as the symmetry of four-dimensional quantum field theories later introduced by Wess and Zumino. The success of quantum chromodynamics in 1973 as a theory of strong interactions put string theory and superstring theory pretty much in the shade as a basis for theories of strongly interacting particles. But almost immediately, Tamiaki Yoneya, and, independently, Joel Scherk and John Schwarz, proposed that string theory should be considered not as a theory of strong interactions but as a theory that unifies all forces of nature, including gravitation. Gravity is the one force of nature for which the Standard Model does not account.

Given this vital attribute, the bringing in of gravity, string theory became a favorite playing field for some of the leading theorists of our time, including Michael Green, Joe Polchinski, John Schwarz, and Ed Witten. I wrote a few papers on string theory, papers of really monumental unimportance. One of them was a lecture at the Third Jerusalem Winter School of Theoretical Physics, which was entitled "Strings and Superstrings." I then stopped my own work on string theory. This was a reflection partly on the theory, and partly on me.

String theory involves mathematical specialties like topology that were not part of my Cornell and Princeton education. I can learn this math when I have to, but I do not come to it as easily as younger theorists who grew up with it. More important, string theory has not led to any new predictions or explanations of quantitative details of the Standard Model that would let us consider it verified experimentally. No progress is made with it. There has been nothing in it like the excitement we felt with quantum field theory in the 1970s.

This lack of progress, or indeed any idea of how to make progress with it, has led some theoretical physicists to disparage further work on string theory. But I tend to disagree with them. String theory has great attractions. It does explain the existence of gravitation, because one of the modes of vibration of a string in any version of string theory is the graviton, the quantum of gravitational radiation. This in itself is a great achievement. Moreover, string theory does not have to be renormalized. It is better than renormalizable. Strings, unlike point particles, are extended objects. So string theory leads to no infinities at all.

In addition to these virtues, in some formulations, string theory leads to the same relations among the three independent couplings of the Standard Model that follow from field theories that unify the strong and electroweak interactions. It is in harmony with the Standard Model.

It may be a long while before string theory becomes an established part of physics. We have seen delays like this before, as for instance with Yang–Mills theory. But I would bet that string theory will be part of the final answer.

We returned to our house in Cambridge in the late spring of 1986. That July, for the first time, I would be giving the summary talk at a "Rochester" Conference on high-energy physics, the 23rd, this time to be held at Berkeley. Traveling to San Francisco for the occasion, we stayed at the Saint Francis Hotel on Union Square. Every weekday from July 16 to July 23, I drove over the Bay Bridge to Berkeley. A huge task confronted me: to summarize thirteen rapporteur's talks, which themselves summarized hundreds of talks in parallel sessions on every aspect of elementary particle physics. As the meeting continued, I worked up a set of transparencies that I could use in my summary talk. On the last day, when I was to give my talk, I left the transparencies at the hotel when I went to Berkeley. Fortunately, I was able to reach Louise by telephone; she took a taxi to the Berkeley campus, and arrived with half an hour to spare, so I was able to give my talk. I began by reminiscing about the 1966 Rochester Conference that had been held at Berkeley, did what I could to summarize the state of the field, and ended with a plea for government funding of new accelerators.

I let it be known that during the 1986–87 academic year, while Elizabeth was at Cambridge, England, I would be very glad to receive invitations to visit there. As it happened, the University of Cambridge Mathematics Faculty had instituted a series of lectures in honor of Paul Dirac, who had died in 1984. Feynman gave the first lecture in the series in the spring of 1986 and I gave the second in the fall.

My lecture was supposed to be at the level of undergraduate physics students. I spoke about what we can now guess about a final theory of physics. Because it is so hard to think of any way of modifying quantum mechanics without producing nonsense, like negative probabilities, I guessed that quantum mechanics would always be part of the story. The experience of twentieth-century physics led me also to guess that principles of symmetry would play a main role. I laid out the general idea of effective field theory, as an approximation at low energies, of what cannot currently be observed at the energies within the capabilities of the accelerator at CERN. The Standard Model can be understood as an effective field theory, in that it might well be a valid approximation to a more fundamental theory, such as string theory, that would be verifiable only at higher energies than we can achieve at present. Presuming this to be so, we can conclude that, since string theory is unconstrained by a requirement of renormalizability, we can work with effective field theories as if similarly unconstrained.

Fortunately, 1987 was the 300th anniversary of the publication of the greatest book in the history of physical science, Newton's *Principia*, and I was invited back to Cambridge, England, to speak at the celebration. This gave Louise and me a second chance to see

Elizabeth there. My talk picked up on testimony I had given before Congressional committees, advocating the funding of a large facility for high-energy physics, the Superconducting Super Collider, more about which in Chapter 14.

This talk was pretty much a reductionist manifesto. I mean by this that theoretical particle physicists could feel themselves engaged in a historic enterprise, dating from Galileo, through Newton and Maxwell and Einstein, to discover, at bottom, the ultimate few, simple, laws of nature. Our theories would be increasingly "beautiful," as I like to think. In my writings and talks, I speak of "beautiful theories" – beautiful because they would contain no arbitrary terms and require no insertions or corrections. One could not change any feature of them. Characteristically, they simplify, clarify, explain, and give broader reach to the successful theories of the past.

In my reductionist thinking, when we discover successful generalizations about nature, we ask why they are true, and we find answers partly in terms of contingencies in the nature of the problem we pose, but often partly also in terms of other generalizations. This gives a sense of direction in science. Principles are explained by deeper principles, but not vice versa. One great example is Newton's explanation of Kepler's laws of planetary motion, that planets move on ellipses with the Sun at a focus, sweeping out equal areas in equal times, and with the squares of their periods proportional to the cubes of the major axes of the ellipses. Newton's laws of motion and gravitation are true (at least as very good approximations), and Kepler's laws of planetary motion are also true, but no one would put them on the same level. There is no doubt in our minds that Newton's laws explain Kepler's, not the other way around. In this Cambridge talk, I went on: "There are arrows of scientific explanation, which thread through the space of all scientific generalizations. Having discovered many of these arrows, we can now look at the pattern that emerges, and we notice a remarkable thing: perhaps the greatest scientific discovery of all. These arrows seem to converge to a common source." This common source, which we do not yet know, is what I call a "final" theory. At present, the search for a final theory takes us into the world of short distances and high energy explored in accelerator laboratories. Of course the search for a final theory is not the only interesting kind of science, or even the only interesting kind of physics, but it has a unique importance of its own, which I think justifies spending large sums on accelerators.

In the summer of 1988, I gave a talk in Stockholm in honor of Oskar Klein, the theorist who had rolled up Kaluza's fifth dimension. This was the only time, apart from Nobel Weeks in December, that Louise and I have spent in Sweden. The Grand Hotel was hosting some sort of conference, so for the only time we stayed in a different Stockholm hotel, the Strand Hotel, also on the water's side. It was nice to see the city in sunshine, which lasted all evening.

It was in the late 1980s that Louise and I started to use computers. The law school gave its faculty desktop work stations attached to a central server, all manufactured by the Wang Corporation. Having learned to use the Wang operating system, Louise and other law faculty bought stand-alone Wang computers to use primarily for word processing in their home offices.

Of course personal computers have been a huge boon to us, for word processing, email, bank account management, and much else. In my first thirty years as a physicist, I had to write physics articles by hand, have them typed by a secretary with space for equations, and then I put in the equations by hand. If I wanted to send out preprints, the whole process had to be repeated on a mimeograph master. Now, along with virtually all physicists, I use a computer program called Latex, about which I learned from Joe Polchinski. It is a word processing program that allows the author to produce mathematical symbols and equations by typing suitable codes. It is an aesthetic pleasure now to create mathematical expressions that will look like those in books, and physics journals generally insist on receiving articles as Latex files. Using this code, one can also send equations by email. Once I was even paid a typesetter's fee from a publisher, because by submitting my manuscript as a Latex input document, I saved them the expense of typesetting it.

In 1988, I published an article for a cookbook. It was a recipe for "Anhydrous Zucchini." It describes how to cook zucchini (courgette, to some) so that, instead of producing the usual watery mess, you turn zucchini into tasty dry slices. The secret, I pointed out, is to keep turning the slices over nervously while they frizzle in olive oil. This was a contribution to a book, *But the Crackling is Superb*, consisting of recipes contributed by fellows and foreign honorary members of the Royal Society of London. I had been proud in 1981 to be elected a foreign honorary member of this society, which Newton had served as president once death had removed his enemy, Robert Hooke, from its meetings. Its distance has kept this society from being very important in my life, although on a visit to London, Louise and I did once stay at the society's handsome house in Carlton House Terrace. As a foreign member, I cannot even vote in an election of new members. But at least with this article, I was able to do my bit for British cuisine.

14 Super Collider Days

After the discovery of the W and Z particles in 1984, there remained one large question about the electroweak theory. Is the electroweak symmetry breakdown that is responsible for the W and Z masses due to the appearance of nonzero values of one or more new fields, as in the original electroweak theory, or to effects of some sort of new strong interaction, as in the technicolor theories of Susskind and myself. In the former case, there would be at least one new weakly interacting particle that came to be known as the Higgs boson, having a mass of one or two hundred proton masses, which would be above the mass scales already explored experimentally. In the latter case, there would be a host of new strongly interacting particles at similar high mass scales, like the familiar mesons and baryons but with much larger masses. In either case, the answer could only be found at an accelerator that could give particles in its beams energies higher than had been so far available.

Of course this was not the only thing that might be learned at a higher-energy accelerator. It might be possible to produce the particle of dark matter, or the squarks and gluinos called for by theories of supersymmetry. Accelerators are sometimes built with no specific target in mind that might be discovered at higher energy. This was true of the Proton Synchrotron at Conseil Européen pour la Recherche Nucléaire (CERN). When it was built, no one guessed that it would lead to the discovery of neutral current weak interactions. But if possible, it is nice to know that there is at least one question that will be settled with a new accelerator.

Two groups of physicists in Europe and America began to lay plans for a high-energy accelerator that could settle the question of electroweak symmetry breaking. Some aspects of these plans were inevitably the same for both groups. It would be necessary to accelerate particles in a ring, so that they could be given more and more energy as they went many times around the ring. The particles accelerated would have to be protons rather than electrons to avoid the huge loss of energy in what is called synchrotron radiation, emitted when a charged particle of low mass like an electron is sent at high energy around a curved path. Intense magnetic fields would be needed to hold the protons in their curved paths, fields that could only be produced by magnets using superconducting cables. Also, if the accelerated protons were to collide with stationary target particles, most of their energy would be wasted in the recoil of the target particle, and it would have to send two beams of protons around the ring in opposite directions, to collide head on, at various stations around the ring, where experimenters could study what is produced in the collisions. Even so, protons are composites of quarks and gluons, and in consequence only a small fraction of the energy in a pair of colliding protons would be available for the creation of new particles. In order to produce the particles with a mass of one or two hundred proton masses associated with the breakdown of the electroweak symmetry, it would be necessary for a pair of colliding protons to have a total energy equivalent to over ten thousand proton masses.

In July 1983, a panel headed by Stanley Wojicki of Stanford recommended that the Department of Energy (DOE) should scrap existing plans for a new accelerator at Brookhaven National Laboratory that would not have had enough energy to settle the question of electroweak symmetry breaking, and should start instead on a proton collider of higher energy. Their recommendation was accepted by the High Energy Physics Advisory Panel of the DOE, which gave the accelerator its name, the Superconducting Super Collider, or SSC. In December, the Secretary of Energy ordered work to stop on the Brookhaven accelerator, and asked Congress for authority to divert its funds to the SSC. Meanwhile, in Europe, plans were laid for a large proton collider at CERN. This would become known as the Large Hadron Collider, or LHC. It would be considerably less powerful than the SSC.

My direct involvement with the SSC began in 1984 and continued to occupy a good part of my life for the following decade. I was on the SSC Board of Overseers from 1984 to 1986. The task of preparing a detailed design was handed to a central design group at Berkeley, headed by Maury Tigner, an ace accelerator and superconductivity expert at Cornell. By April 1986, the design was complete. The SSC would be housed in a ten-foot-wide underground tunnel forming a fifty-two-mile-long oval ring. Inside the ring, two beams of protons traveling in opposite directions would each be accelerated to an energy equivalent of about twenty trillion proton masses. The protons would be kept on track and focused by 4,728 superconducting magnets. Since superconductivity requires very low temperature, the magnets would be kept cool by about 2 million quarts of liquid helium, held at about 4 degrees Celsius above absolute zero (4 Kelvin).

The Reagan administration approved the project in January 1987, and with other physicists I was sent by the DOE to the White House to thank President Reagan for this decision. It was an odd experience. In my time, I have met many politicians: four presidents, numerous members of Congress, and countless politicians here in Texas, all while politicking for arms control and the SSC. One skill that all these politicians share is an ability to make contact, to give an impression (true or not) that they are interested in the person introduced to them. All except Reagan. In shaking hands with him and thanking him for his SSC decision, I had a feeling that there was no one there. I wonder if his Alzheimer's disease had begun much earlier than when it was reported publicly in 1994.

On April 7, 1987, I joined several other physicists testifying in favor of the SSC project before the Subcommittee on Energy, Research and Development, of the House Committee on Space, Science, and Technology, and the Subcommittee on Energy, Research, and Development, of the Senate Committee on Energy and Natural Resources. (This is the testimony to which I referred in my talk at the celebration of Newton's *Principia* at the University of Cambridge, mentioned in Chapter 13.)

FIGURE 14.1 White House Rose Garden ceremony for the SSC. Secretary of Energy John Herrington is at the podium; President Reagan is flanked by the author (left) and James Cronin (right); Samuel Ting and Burton Richter are in the back row

One of the senators who heard my testimony remarked that, at the moment, the project had the support of 100 senators, but when a specific site was selected, the number supporting the SSC would drop to two. I did not fully take this in. At the time, I did not appreciate the importance of what is commonly known as "pork." For the success of the SSC, it probably would have been helpful to emphasize to the Senate that, in the construction and running of the SSC, a "buy American" rule could be imposed, and contracts could be let in a large number of states for the SSC's various components and requirements.

In April 1987, the DOE began a site-selection process. The states were invited to submit proposals for specific sites, to be judged according to the suitability of local geology for the underground tunneling and the resultant tunnels that would house the accelerator beams; the availability of nearby infrastructure, including power lines, roads, and airports; acceptability of the project by local residents; possible environmental impact; and amenities that would help to attract permanent staff. The deadline for these proposals was set as September 2, 1987.

I then became involved in the site-selection process. The first step was a competition within each interested state to choose the one or two sites to propose to the DOE. In 1987, I served on the Texas SSC High Energy Research Facility Advisory Council that would deal with this choice. The word "advisory" was no joke; it was made clear to us that the final decision would be made by elected politicians, not by scientists.

Pursuing our own initiative, Frank McBee and I went to the Chamber of Commerce in Austin to urge them to work up a proposal for a site east of Austin, in a geological formation called the Austin Chalk, particularly suitable for tunneling. Other sites were proposed, including one in West Texas and one in the part of the Austin Chalk southwest of Dallas near Waxahachie. Our advisory committee decided that the best sites were the ones near Austin and Waxahachie, but at a higher level, it was decided instead to propose the West Texas and Waxahachie sites to the DOE. The Waxahachie site was a good choice, but the West Texas site had nothing to recommend it, not in geology, infrastructure, or amenities. It was explained to me that it was necessary to include some site in West Texas to preserve statewide support for the project. People in West Texas tend to be suspicious of cities to the east, especially Dallas and Austin. It was a good lesson in practical politics.

Then I served in 1988 on the SSC Site Evaluation Committee of the National Academy of Sciences and the National Academy of Engineering. We received forty-three proposals of possible sites, weighing altogether about 3 tons. I still have these proposals in boxes under a table in my Texas office; it was too much trouble to dispose of them. Some proposals were hopeless. When committee member Murph Goldberger saw the West Texas site proposal, he called it "bizarre." There was a site proposed near the northern border of New York State that was so far from any American airport that scientists working at that site would have to fly in to Montreal. There was a site in Arizona that was considered to be safe from any environmental impact from the SSC because it was already so highly polluted. I urged the committee to include among desirable amenities the availability of good guacamole, but was voted down. One practical suggestion I made that was accepted was to judge the size of airports by measuring the height of the columns listing incoming flights in the Official Airlines Guide.

In the end, our site evaluation committee sent a list of seven "best qualified" sites to the DOE. Two of them were clearly superior. One was the Texas site near Waxahachie, which had the benefit of excellent geology, the very large nearby Dallas–Fort Worth airport, and enthusiastic local support. The other was near Fermilab in Illinois, which had only fair geology and faced some local opposition, but was accessible from the even larger O'Hare airport and would have had the great advantage of being near America's largest existing accelerator laboratory.

Around this time, Leon Lederman, the director of Fermilab, made a trip to Waxahachie to spy out the competition. As he later described it in a talk in Austin, he knew that the name of the town is difficult to pronounce. It is not Waxahachie, but something more like "Wuhxahachie." So in a fast-food restaurant in Waxahachie, he asked the woman behind the counter, "How do you pronounce the name of this place?" She looked puzzled and replied, with slow emphasis, "Why, honey, it's B-u-r-g-e-r K-i-n-g."

On November 10, 1988, the Secretary of Energy announced that the SSC would be located in Texas at the Waxahachie site. This was big news in Texas. Governor William Clements convened a joint session of the Texas House and Senate to celebrate. After the session, I found myself standing in the House chamber next to the Governor. When he said to me how glad he was about this, I replied that I just hoped that Congress would continue to fund the project. Clements told me not to worry; Jim Wright would take care of it. The Texan congressman Jim Wright was the powerful new Speaker of the U.S. House of Representatives. Clements may have been right that Wright would have been able to protect funding for the SSC, but an ethics investigation into payments to him and his wife led Wright to resign from Congress in July 1989.

In 1991, I received a heartbreaking email message from my oldest friend in physics, Gary Feinberg. He wrote, "I am dying, old friend." Gary died in 1992. I do not know the circumstances, but his early death was not only tragic but also ironic. He had interested himself in a cryogenic theory that a dying human could be frozen and brought back to life when the cure for what ailed him had been found. He persistently denied himself animal protein and other food, believing that a low-calorie diet saved the energies of the body and prolonged life. I early suspected that the resultant almost skeletal slenderness he achieved was destructive. A typical lunch meeting with Gary would find him ordering a sandwich of lettuce and tomatoes on toast. Yet nutritional science was sufficiently advanced at the time to have guided him to a more nourishing diet. He died of cancer. I hope his death was unrelated to his self-imposed dietary restrictions.

From 1989 to 1993, I served on the Science Policy Committee of the SSC Laboratory. We met in a temporary office building in De Soto, a town between Dallas and Waxahachie. In January 1989, the experimental physicist Roy Schwitters was recruited as director of the laboratory. I thought that this was a great choice. Roy had been a key member of the experimental team at Stanford Linear Accelerator Center headed by Burt Richter. Among other things, that team had discovered the *J* particle. Roy had then become a leader of a major experimental collaboration, the Collider Detector at Fermilab (CDF) collaboration.

Louise and I had come to know Roy and his wife Karen after his move from Stanford to Harvard, and we liked them very much. Not only was Roy a superb experimentalist, he seemed to me to have an ability, shared with Murph Goldberger, to give direction to groups of fractious physicists, not so much as a commander but as a primus inter pares, a first among equals. As part of the inducement to Roy to leave Harvard and come to Texas, the University of Texas at Austin agreed that, when his time as SSC director was over, he would be offered a professorship in our physics department.

Schwitters seemed to enjoy the move to Texas. He became friends with Dallas leaders who supported the SSC program, including high-tech CEO Morton Myerson, who sponsored a fine new concert hall for the Dallas Symphony, and Margaret McDermott, widow of the founder of Texas Instruments.

In September 1989, Congress appropriated $276 million for the SSC, with part of the appropriation for the first time designated for construction. This was good news. Research and development can go on forever, but construction needs to follow a definite schedule. Nevertheless, every year, physicists had to struggle to keep the project funded.

For many congressmen, the only thing that mattered in their view of the SSC was the prospect of its direct benefit to their constituents. One of the strongest advocates of the SSC was Joe Barton, who

FIGURE 14.2 The author giving a presentation on the SCC in Dallas in 1989

represented the Texas congressional district where the SSC would be built. I never saw in him the slightest interest in what might be discovered at the SSC. On the other hand, some members of Congress were alienated by the fact that their state or district had not been chosen as the SSC site. For instance, Sherwood Boehlert was the congressman whose district included the site near the Canadian border that had been proposed by New York State. He had at first been an advocate for the SSC, but turned against it when the site in his district was not included in the short list considered by the DOE. Once when I was walking through a tunnel under the Capitol along with a group of physicists who had been lobbying for the SSC, we encountered Congressman Boehlert coming the other way. One of our groups handed him a button saying something like "Fund the SSC." Boehlert handed it back, saying we should keep it, because it would become a historical relic.

The choice of Texas was particularly annoying to some members of Congress. They suspected that the SSC might be just another Texas pork barrel project. It did not help that at just that time Texas was also strongly supporting the development of the International Space Station, to be administered by the Johnson Manned Spacecraft Center in Houston. At one hearing on the SSC by a Congressional committee, a member of the committee commented that he understood how the International Space Station would help us to understand the universe, but he could not see that in the SSC. I could have cried. As most scientists understood, the International Space Station would do nothing for science that could not be accomplished at much less cost by unmanned satellites. In contrast, the SSC would reveal things about matter that would be relevant to our understanding of the early universe, and might reveal the nature of the dark matter particles that dominate the universe now. But I had already given my testimony, so I had no chance to reply to the congressman's comment.

In its competition for congressional support with programs of manned space flight like the International Space Station, the SSC might have been hurt by the fact that it was so much less expensive.

The International Space Station would cost roughly ten times what the SSC would have cost, and spread its funds to subcontractors throughout American states far more widely than the SSC ever could.

There was a deeper issue underlying the debate over the SSC. Elementary particles are studied at accelerator laboratories only as a means to an end, the discovery of the fundamental physical principles of science that govern nature. This is not the only interesting kind of science, or even the only interesting kind of physics. There are lots of important and interesting problems that have resisted solution even though we know all we need to know about the relevant underlying physical principles, because the application of these principles has proved too difficult. In this class of "inner" physical problems I would list high-temperature superconductivity, galactic spiral arms, energy distribution in turbulence, and so on. This is in contrast with frontier problems of physics. For instance, why do the masses of the quarks and leptons take the values we observe? What is the dark matter? What describes nature at the very high mass scales where gravitation is as strong as the other forces? What was going on in the very early universe? And so on. These are the problems we hope to address using high-energy accelerators, large telescopes, and other expensive facilities.

Some members of Congress did not understand this. At one committee hearing, a congressman asked why, instead of spending billions on the SSC, we did not spend much less on a large computer that could just calculate what would be observed in collisions at the SSC. He did not understand that calculations like this are impossible because we do not know the principles that govern the phenomena that would occur when the SSC takes us to a higher level of energies, and in any case it is not the collisions that interest us, but the principles that govern them.

Other members of Congress did not care. Once I was on the Larry King radio show with a telephone link to a congressman who opposed the SSC. He said that he was not against spending on scientific research, but that it was important to set priorities. I replied that the SSC would help us learn the laws of nature, and asked if that did not deserve a high priority. He answered with a single word: "No."

This conversation mirrored a dispute that was going on among physicists. There was no argument among elementary particle physicists about whether the SSC should be built, and no argument among physicists in general about whether it would do good research. But some physicists who did not work on elementary particles thought that the money to be spent on the SSC would be better spent on research on other kinds of physics, such as their own. Two of them had testified against building the SSC at the hearings held in Congress in 1987. Both worked on condensed matter physics, the study of solids and liquids, including semiconductors, superconductors, and superfluids.

One of these SSC opponents was James Krumhansl, who had been one of my teachers at Cornell. His testimony was particularly damaging because that year he was president-elect of the American Physical Society. Meeting by accident at the Washington airport before the hearings, I told him that I thought that his testimony against the SSC would be improper, because, although he had every right to express his own views in this case, they would be seen by members of Congress as the judgment of the Physical Society, which they were not. He shrugged this off.

The other physicist opposing the SSC was Philip Anderson, whom I regard as being at the time the leading condensed matter physicist in the world. In 1972, Anderson was the author of an influential article "More is Different." In it, he explained the idea of "emergence," that from the elementary constituents of the world, there emerge phenomena that cannot usefully be described in terms of their constituents. There is nothing like emotion on the level of individual living cells, and nothing like temperature or pressure on the level of atoms or molecules. I agreed with this, but drew a different conclusion, that whether or not it will help us to understand emergent phenomena, there is a special interest in discovering the underlying principles from which they emerge. I never convinced Anderson of this, but he was so punctilious in his testimony at acknowledging that the SSC if funded would be successfully built and would do worthwhile research that his testimony may have done more good than harm.

I was glad to see that some congressmen did understand the aims of research in elementary particle physics, and agreed with them. One was Jerrold Nadler, who represented the 10th congressional district of New York, located on the east side of Manhattan. I doubt if a penny of SSC funding would ever have found its way to Nadler's district, but he gave an eloquent speech on the floor of the House in favor of funding for the SSC. Another was Senator Bennett Johnston of Louisiana. His state would supply some of the magnets for the SSC, but my conversations with him convinced me that he was seriously interested in research at the frontier of physics. At one point, though, I made a mistake with him. We were talking about Senator Dale Bumpers of Arkansas, an active opponent of spending on the SSC, and I remarked that I thought Bumpers was not very bright. Johnston froze. He and I were allies on the SSC issue, and Bumpers was an adversary, but he did not want to hear criticism of a fellow senator from a nonsenator. I should have known better. Ordinarily I take care not to speak ill of others. I did not repeat this mistake.

As time passed, it became clear that the SSC would cost considerably more than the $4.4 billion originally projected. This was in part because the $4.4 billion was supposed to cover only construction of the accelerator, not other necessary items like the detectors that would sort out what happens in the head-on collisions of protons (some of which would be paid for by laboratories abroad). All this was spelled out in the original proposal, but changes in the definition of project cost made it look like the cost was increasing. Also, the cost *was* increasing, because Congress was not maintaining spending at the originally supposed rate. Spreading out the time for construction meant that more would be spent, in part because it would be spent in depreciated dollars. And there had also been necessary changes in the SSC design that increased its cost. Our Science Policy Committee received advice from accelerator experts that it would be necessary to increase the aperture in the superconducting magnets through which the protons would pass, in order to avoid the risk of having protons hit the magnets, quenching their superconductivity. We knew that the considerable increase in cost that this would entail would hurt the SSC politically, but we thought that building an accelerator that did not work would be worse than not building it at all. The aperture increase was approved.

The case for the SSC was hurt by comments from supporters of its competitor, the LHC at CERN in Switzerland. They claimed that the LHC would be much less expensive than the SSC because it used an existing tunnel, which had housed the Large Electron–Positron Collider. The circumference of this circular tunnel was considerably smaller than that planned for the SSC, so it would not be possible to accelerate protons at the LHC to an energy beyond about a third of the energy planned for the SSC, but LHC boosters claimed they could make up the difference in energy by running at higher intensity. This argument was disingenuous. It is true that by running at higher intensity the LHC could make up some of the advantage in the number of interesting collisions per second that higher energy would give the SSC. But high intensity comes at a cost. When the detectors at the LHC would find that an interesting collision had occurred, as a result of its high intensity, there would be several collisions going on at the same time, and it would be necessary to sort out what particles were being produced at which collision. And, of course, the LHC's lower energy reduces the range of masses of new particles that it can produce. The SSC could discover squarks or weakly interacting massive particles or other particles that the LHC would not have enough energy to create.

Apart from these real issues, there was nonsense in the press about runaway spending at the SSC. One article reported reckless expenditures on potted plants at the temporary SSC headquarters in De Soto. In fact, these very small costs were covered by a fund that had been specially granted to the SSC director to use for amenities that would brighten employees' working experience. There was no reckless spending in the SSC program.

The *New York Times* was hostile to the SSC, perhaps because of the project's association with President Reagan, or with right-wing Texas. I joined some physicists in an interesting meeting with the *New York Times* editorial board. We explained the process of building the SSC and its aims, after which the hostile treatment of the SSC in the *Times* ceased, though the paper never actively supported the program.

In 1992, the House of Representatives voted to delete spending for the SSC from the energy and water appropriations bill. SSC funding was supported in the Senate, and was restored after a favorable report by the House–Senate conference committee. Though funding would continue for the present, it was clear that the future of the SSC was not secure.

That year, with several other physicists, I was invited to the Executive Office Building to discuss the SSC with the new vice president, Albert Gore. He assured us that the Clinton administration would not only support the SSC, but would do so actively. I was not convinced, but Gore won my heart when, on his way out, he paused, came back into the room, and said that he did not want to miss this opportunity to ask a question that had bothered him: What was there before the Big Bang? I was the only physicist in the room who had worked on cosmology, so I was elected to answer. I gave the usual unsatisfying reply, that we don't know if there was a beginning rather than an eternal past, but if there was a beginning then presumably it was the start not only of matter but also of time, so there was no "before." Such equivocal and Zen-like answers are inevitable, but do not improve relations between physicists and other people.

I made a brief trip to Tokyo the fall of 1991, on behalf of the SSC, to no avail. I was up all night, and with the days and nights turned upside down, out of my hotel window I saw all the Americans out running at 4:00 a.m.

During 1993 and 1994, I was on Larry King's radio show trying to promote the SSC; I testified in both the Senate (1992) and the House (1993), met with the *New York Times* editorial board, and dined with Senators Kerry and Shelton Smith.

As it turned out, the Clinton administration did go on supporting the SSC, but not very fervently. There was a story, the truth of which I have not been able to establish, that the administration told Texas Governor Ann Richards that it could not actively support both of the two big Texas projects, the SSC and the International Space Station, and she opted for the Space Station.

Once again, on June 24, 1993, the House of Representatives voted to remove funding for the SSC from the energy and water appropriations. As in 1992, the Senate voted in favor of SSC funding, and once again the House–Senate conference committee voted to restore the funding. Our Austin congressman, Jake Pickle, told me at the Headliners Club that this time the House would take this as an insult. So it seemed. On October 19, 1993, the House voted to reject the conference committee report and send the appropriations bill back to the committee with SSC funding again deleted. The committee agreed, and the SSC was dead.

What a waste. About a billion dollars had been spent in digging the SSC tunnel and preparing to manufacture superconducting magnets. When I visited the SSC site in the autumn of 1991, I saw many boarded-up farm houses whose owners had been forced to leave. Some physicists had spent a good part of their careers helping to design the SSC and preparing to use it.

It is to the credit of the Space Station that it could service the invaluable Hubble Space Telescope. But for the same money, we could have had some several duplicate Hubbles, which would have made repairs to one of them unnecessary. Still, surprises were in store for us in this golden age of cosmology. By 1996, astronomers were assured they could seriously lay plans for the next generation of space telescopes, a space telescope that would have far more reach and power than Hubble. It has shaped up to have a multi-mirrored hexagonal lens of enormous size, strongly influenced by the proven advantages of the Hobby–Eberly mirror here in Texas, affiliated with the McDonald Observatory. It is one of the ambitions of my latter days to live to see the launch of the James Webb.

Tragedies sometimes come with side benefits. The University of Texas at Austin reaped the great benefit of recruiting Roy Schwitters to its faculty.

For me, the demise of the SSC was not without other consolation. I had spent so much effort in explaining the reductionist aims of elementary particle physics that I wrote a book about it, *Dreams of a Final Theory*. This was to be a trade book, rather than a publication of the academic press, and for the first time in my life I needed a literary agent. Richard Rhodes, our tenant at One Berkeley Place, suggested Morton Janklow, his own agent. Morton is a leading agent who, besides Rhodes, has represented popular novelists like Anne Rice and Danielle Steel, as well as four US presidents. Perhaps because he had not represented any physicists, he agreed to act as an agent for *Dreams of a Final Theory*. As it happened, my book was published in 1992, and made pots of money for both Janklow and me. This was, alas, not in consequence of the special virtues of my book.

Stephen Hawking's book *A Brief History of Time* had come out in 1988 and had become a runaway bestseller. Publishers in the US and abroad were somehow convinced that *Dreams of a Final Theory* would be the next big thing in popular science books, and offered huge advances. My book got good reviews, and did sell well, though nowhere near as well as Hawking's. Fortunately, unjustified advances on royalties do not have to be returned. This book went far to helping Louise and me buy our present wonderful house on Lake Austin.

At not much less total cost than would have been spent on the SSC, CERN did continue with the LHC. In 2008 it started operation, and in July 2012 enjoyed its crowning success in discovering the Higgs boson. Thus clicked into place the final prediction of the Weinberg–Salam electroweak theory, settling the question raised at the beginning of this chapter: whether or not the symmetry underlying the electroweak theory is broken by the appearance of nonzero values of new fields, as in the original electroweak theory. Since then, much effort has been expended in studying the properties of the Higgs boson, and so far they seem just what was predicted in 1967–68 by the simplest version of the electroweak theory, chosen for its very simplicity as the easiest case.

It is not important in my work whether experimental data like this comes from Europe or America. The data is all made open to all physicists. As an American, I am sorry nevertheless to see that our country is no longer in the forefront of experimental research on elementary particles. One consequence is that American students will either not work on experimental high-energy physics or, if they do, will tend to study and work abroad. Our young theorists may find this a more convenient arrangement for themselves as well. We are thus losing the sort of cadre of scientists that has served America well in the past. The good theorists were here for our country when the national defense required them to be at Los Alamos. Will they be here, or abroad, should the national defense again have need of them?

I am heartened by steady confirmation of my thinking at the LHC. But, as a physicist engaged in our larger project, I have some regrets. As expected, the LHC has reached only about a third of the energy that would have been available at the SSC. It has discovered the Higgs boson, as predicted, but nothing else exciting. Dark matter, or signs of supersymmetry, or something else completely unexpected, might have been found at the SSC, and now will not be found for decades to come.

15 Austin: The 1980s

Louise's comic stories about my morning showers and the cosmological constant, with me chronically stepping out of the shower in uncertain conviction or certain despair, are more or less true. This problem, even as it changed over time, has teased the wits of a good many theorists. In May 1988, I returned to Harvard, two decades after my first Loeb lectures, to give a second series of Loeb lectures on "The Cosmological Constant Problem." The series might have degenerated into another stepping-out-of-the-shower exercise in inconclusiveness, but, for better or worse, it did not. Its thesis, controversial as it is and was, did eventually become a matter of some interest.

In these lectures of the 1980s, I surveyed a number of proposed solutions, and found difficulties with each of them. I did outline a theorem that usefully ruled out a large class of proposed solutions. Then I proposed what I thought was the only solution that works. I was very proud of this achievement, although it remains controversial because it is not purely a consequence of quantum field theory but at bottom relies on simple reasoning.

In 1987, I had published a brief paper, "Anthropic Bound on the Cosmological Constant." Of course Einstein's posit of "Λ" was essentially anthropic. It is what is needed to keep the universe from collapsing in gravitational catastrophe, or escaping from itself, since it is expanding. It is expanding, but keeping together as it expands. And "Λ" is what accounts for this.

I began it with a review of various fashionable cosmological theories that lead to a multiverse of many parts, one of which is our observed Big Bang. It is an attractive theory because one can posit a steady state of constant creation of universes, thus ridding ourselves of the philosophically unpalatable singular Big Bang.

Andrei Linde had considered a universe pervaded by fluctuating fields. In countless regions in which a field happens to be particularly large, that region undergoes a rapid expansion to a very large size, like our own Big Bang.

Stephen Hawking had a theory based on the peculiar features of quantum mechanics that led to a similar multiverse. It was expected in these theories that some of what we call constants of nature, the fundamental laws of nature, would be different in different parts of the multiverse.

I recall another story that Louise tells, that she confided this conclusion – that there might be different sets of laws of nature – to her dinner companion one evening, the great condensed-matter theorist Philip Anderson, and that Anderson comfortably agreed with her own view, insisting, "Louise, there is only one set." And it is true that the theories I am discussing now contemplate differing laws of nature among the universes in a multiverse. But Louise is right to be shocked by the suggestion, and may be right to stand by Phil Anderson's view that there is only one set. Indeed, my lifelong project, my reductionist program, in which Louise was thoroughly indoctrinated, could not yield an ugly solution such as an increasingly complex set of solutions. It is not what I would have called a "beautiful" theory.

With my anthropic principle, I felt and still do feel that I had come to the rescue. It is true that my resolution can accommodate a multiverse. But it could also describe the singularity of our own universe.

What my anthropic theory boils down to is this: Any scientist who measures the constants of nature will have to find values that, however rare in the multiverse, are consistent with the emergence and sustaining of life, and with its evolution into beings capable of science. This is known as the anthropic principle. It is a specific version of an ancient idea. The Roman physician, Galen, argued that the gods put the Sun at just the right distance so that the Earth would not be too cold or too hot for the existence of life. Modern religions think of this as the "intelligent design" of the Creator, and argue that it could not be accidental. And yet of course the conditions of stars and their planets are of infinite variety and are historical accidents, not at all the product of design. In physical theories, whether or not they lead to a multiverse, the anthropic principle is just the constraint of common sense, with no need to invoke benevolent gods.

Indeed, in my 1987 paper, I noted that, unlike other constants, the cosmological constant can affect the appearance of life in only one way. If the magnitude of the dark energy, including field fluctuations and the cosmological constant, were much larger than the cosmic density of the mc^2 energy in ordinary matter, it would prevent the formation of galaxies, and hence of stars and planets on which life could arise. If it were too large, the galaxies would fall apart. If it were too small, they would implode. In either case, human life as we know it could not exist. Anthropic reasoning thus could make it possible for us to calculate the effective vacuum energy. The cosmological constant is that which makes it possible for us to be here asking the question.

My 1987 calculation of the anthropic upper bound on the vacuum energy density had been based on rough estimates of the initial inhomogeneities that would condense into galaxies if the condensation were not prevented by the repulsive forces produced by a cosmological constant. But in the late 1990s, two teams of astronomers were carrying out new, highly detailed studies of the expansion of the universe, and it was necessary to give a more precise anthropic upper bound. In 1997, I joined forces with my friend Paul Shapiro, a colleague in the Texas astronomy department, and postdoc Hugo Martel to do a more precise calculation of probabilities that a random astronomer anywhere in the multiverse would observe various values of the dark energy density. Our calculations showed that it was unlikely that the dark energy density would be greater than about nine times the present cosmic mass density, and very unlikely that the dark energy density would be smaller than the present mass density. We had trouble getting this article published in the *Astrophysical Journal*. The editor, like many other scientists, was not in sympathy with anthropic reasoning. We overcame her objections by pointing out that, if, as widely expected, observations were to show that the dark energy density is in fact less than the cosmic mass density, then our calculations could be cited as evidence that the anthropic principle could not explain the smallness of the dark energy.

Our paper was accepted after eight months, and published in 1998. A little later that year, dark energy was discovered in independent studies of the recession velocities and distances of galaxies, by two groups of astronomers, the Supernova Cosmology Project, headed by Saul Perlmutter, and the High-z Supernova Search Team, headed by Adam Riess. (Both groups used measurements of the apparent brightness of the exploding stars that are known as type Ia supernovae (read: "type one-A") to calculate the distances of the galaxies that contain them.)

It was these supernovae observations that also gave us the new information that the universe is not only expanding, which we knew, but that the expansion of the universe is accelerating. The acceleration was understood as probably related to the dark energy. By 1998, astronomers were convinced of all this, when the acceleration of the universe was apparently confirmed by both teams. Their data rule out a zero value for dark energy density. Really rule it out – astronomers are confident about these results. This suggests a value about three times the cosmic matter density.

This observational result was not inconsistent with the notion of a possible multiverse. The cosmological constant – the laws of nature – might differ among the various Big Bangs in a multiverse, in theory. Anthropic reasoning cannot prevent unpredictable variety. It is by pure accident that we exist. The only laws of nature that explain human existence are Darwin's.

There has been an ongoing argument, not about whether multiverse cosmology has been confirmed, which no one claims, but about whether it should be taken seriously at all. I gave the opening talk at a conference to discuss this matter, held in the Master's Lodge at Trinity College, Cambridge, in September 2005. From a portrait on the wall, Isaac Newton glowered down disapprovingly. In my talk, I mentioned that I had seen a report that in an earlier conference the astrophysicist Lord Martin Rees (Master of Trinity) had said that he was sufficiently confident about the multiverse to bet his dog's life on it, while Andrei Linde had said he would bet his own life. I bravely raised the ante, and said that I would bet the lives of both Andrei Linde and Martin Rees' dog.

The issue has not been settled. It will be possible to get a clue to the validity of the anthropic explanation of the value of the dark energy density through studies of the cosmic expansion. Such studies can reveal whether the dark energy is constant or is changing with time as the universe evolves. It might be that the dark energy is somehow related to the density of the energy in ordinary and dark matter, so that they are always comparable in magnitude. In some way, any theory must be consistent with our continuing existence in this universe. So far, no sign has been discovered of any change in the dark energy density.

16 The Dark Energy

In December 1994, having been elected President of the Philosophical Society of Texas, I produced a program on "Cosmology." I had always been interested in cosmology. With Louise's gift to me of one of Chandrasekhar's books, I had been turning increasingly to cosmology as a way forward for particle physics. Chandra had explained, as a matter of particle physics, what made the stars shine. But I saw particle physics as a way of explaining the very existence of stars and galaxies in the universe.

Einstein was surely the greatest scientific mind of the twentieth century, and I thought this is where Einstein would have gone had he not engaged his mind in space-time. My mind had often been teased by the question: What is the cosmological constant?

Above all, what was interesting to me was the question whether our understanding of the cosmos could be advanced by particle physics. My first book would be a treatise, *Gravitation and Cosmology* (1972). My intuition was that the answers needed to be sought in the very early history of the universe.

What had especially intrigued me was the three-degree background radiation. The reader may recall from these pages that Penzias and Wilson had discovered this background radiation at Bell Labs. What did it mean? Also, we were teased by the astronomers' discoveries about the very distant stars – new kinds of stars. Light travels at a definite speed, and is limited by that speed. So seeing far away in a telescope is also seeing very long ago, as long as it took light from that distant object to reach us. Astronomers were discovering stars that may not be there now but were there 13 billion years ago, near the time of the Big Bang. That was how long it had taken the light from those stars to reach us. They were different then because they were in an earlier phase of development.

But there was a third impulse behind my turn to cosmology. In part, my turn toward cosmology took place because I was disheartened. There would be no Superconducting Super Collider (SSC) to which one could look for corroboration of the Standard Model. The SSC was dead. I had no confidence in the much smaller and cheaper Large Hadron Collider (LHC) at Conseil Européen pour la Recherche Nucléaire (CERN). In this, fortunately, I turned out to be mistaken. Piece after piece of the puzzle would fall into place – amid tremendous excitement. The Super Proton Synchrotron at CERN would discover the W particle! And then the Z! And the LHC would at last discover the Higgs boson! All exactly as the theory predicted.

But I was disheartened anyway.

Accelerators were our big, expensive toy, and Santa Claus did not mean to bring us a more expensive one, ever. But bigger and better astronomical telescopes were being built, and soon the Hubble would be there, in outer space, gathering information beyond our imaginings.

So it seemed to me that quantum field theory might have a useful role to play in examining the early universe, which we were just beginning to see. I became wrapped up in cosmology.

As 1998 drew to a close, I was drawing to a close a line of work in cosmology that is the most controversial work with which I have been associated. It had to do with the cosmological constant. To explain, I have to go back to 1917, the year after Einstein developed general relativity, after 1915–16. In 1917, Einstein turned his attention to cosmology. Like most physicists at the time, he thought that, on average, the universe is static, with no large-scale expansion or contraction. But in general relativity, just as in Newtonian physics, gravitation would tend to draw matter together, into a cosmic contraction. All the galaxies we see in the heavens should be rushing toward us, as gravity shrinks the universe's matter into a more and more dense condition. Yet this was not the way the heavens were looking. So Einstein realized there must be some sort of counterforce.

Necessarily assuming the existence of this counterforce, Einstein modified the equations that govern the gravitational field by adding a term that he called the cosmological constant. This is the energy of something, perhaps something about space and time, that keeps the universe together.

I have to admit that the identification of the cosmological constant was not a mystery that popped into my head in the late 1990s. As recounted earlier, Louise was fond of telling people that, from the earliest days of our marriage, on a typical morning, emerging from my shower, I would have a dazed expression on my face, and say, "Honey, I think I have figured out the cosmological constant!" The next day I would come out of the shower with the same dazed expression, and say, "No . . . ."

As for Einstein's positing of a cosmological constant to begin with – though Einstein seems not to have realized this – it did not quite solve the problem of the mysterious coherence of the universe. His model universe was too unstable for any such solution. The effect of gravitation obviously decreases with an increase in the distance between objects. But the cosmological constant is supposed to act like a repulsive force preventing universal collapse. So it would also increase with distance – it would have to. If Einstein's universe expanded a little, the cosmological constant could overwhelm gravitation, and the universe would go on expanding faster and faster. At the time, we believed this wasn't happening. Similarly, if the universe started to contract, gravitation would overwhelm the cosmological constant, and the universe would contract faster and faster. But this surely was not happening.

As it turned out, these sorts of concerns were misdirected. Astronomical observation of distant galaxies, chiefly by Edwin Hubble at the Mt. Wilson Observatory, revealed in the 1920s that the universe is not static, but expanding. Einstein came to regard his introduction of the cosmological constant as a mistake, since he had used it to keep his described universe stable, in statu quo. But astronomers could be comfortable with an expanding universe, because the expansion could be attributed to some initial explosion, which would impart velocity to the matter being exploded. There would be no need for a cosmological constant at all.

For atheists like myself, this positing of an initial explosion, a Big Bang, as Fred Hoyle derogatively called it back in 1949, was philosophically distasteful. It meant there really was a Beginning. I have already mentioned in this book, in Chapter 8, my little quarrel with Hoyle, who felt some form of a steady-state universe had to be part of our theoretical understandings, since we knew religion was bosh. There was no creator and no creation, as far as Hoyle was concerned.

But taking the Big Bang as a given, as astronomers increasingly felt compelled to do, there was uncertainty about what was really happening in the early universe. In modern quantum field theories, like quantum electrodynamics, or in the Standard Model, fields are continually fluctuating, even in supposedly empty space. This by itself would give the vacuum an enormous energy per volume. Depending on its sign, this could put the universe into a catastrophically fast expansion or contraction. Even if we included only those fluctuations, the wavelengths of which are no smaller than those probed in today's accelerator laboratories, the rate of expansion or contraction would exceed what is allowed by astronomical observation, exceed it by 56 powers of 10. If you are looking for a cosmological constant, this would have to be it.

So we do need to posit a cosmological constant. This constant must behave just like a vacuum energy density, and can be adjusted to cancel the energy in fluctuating fields. But to avoid a conflict with astronomical observation, the cancelation would have to be exact to fifty-six decimal places.

Today the effective vacuum energy density, including the cosmological constant, has come to be called "the dark energy." It is whatever is out there that keeps the galaxies not only falling in toward each other, but in fact rushing away from each other. It is an acceleration. And astronomers are now finding that, in fact, the galaxies are speeding up. The expansion is accelerating.

Yet despite this vast mighty outward increasing push, the dark energy is incredibly small, when you compare it to the energy densities encountered, for instance, in ordinary matter. This is today's problem of the cosmological constant. It has not gone away. What is keeping the galaxies all together even as they rush apart? They rush apart all together.

17 Austin: The 1990s

In the autumn of 1990, Columbia University held a celebration, and invited all Nobel laureates who had studied or worked at Columbia to attend and receive honorary doctoral degrees. It was a little odd, honoring people because they had already been honored elsewhere, but of course it was a well deserved celebration of Columbia.

Sometime around 1990, I received a letter from Clifton Fadiman. Fadiman had been in the business of public intellectual when I was a boy. His was a voice out of the past. Hearing from him was like receiving a letter from the Lone Ranger. Fadiman invited me to contribute to a book to be called *Living Philosophies*. It would contain essays describing the personal philosophies of an interesting set of authors. The list included some good friends of mine: Freeman Dyson, Stephen J. Gould, and E. O. Wilson. There was some satisfaction in being included, with other scientists, among the public intellectuals that fill this anthology's pages.

In February 1991, I received the Madison Medal from Princeton University. It was pleasant to return to the scene of my graduate studies, and to see old friends. At the award ceremony, a dean introduced me by reading the recommendations that I had received from Cornell professors when I applied to Princeton for graduate work in 1955. I was disappointed that, although the letters were favorable enough to get me admitted to Princeton, I had not earned recommendations quite as ecstatic as I would have liked.

One day in the summer of 1991, I received a telephone call from Allan Bromley, the science advisor to President George H. W. Bush. I was to receive the National Medal of Science. I was surprised, but of course happy to be awarded such an honor. The National Medal of Science is awarded yearly by the President to some dozen people in all fields of science and engineering. The ceremony that year was held outside, in the Rose Garden. When I was called up to receive my medal from President Bush, our historic conversation was an exchange of commiserations about the heat and the humidity.

In December 1991, the Nobel Foundation celebrated the 90th anniversary of the first Nobel Prizes, and the Foundation invited back to Stockholm all laureates of previous years (aside from the Peace Prize winners, who went to Oslo).

Once again in the summer of 1992, I gave the summary talk at a "Rochester" Conference on high-energy nuclear physics, this time at Southern Methodist University in Dallas.

In 1992, I was delighted by a chance to visit Salamanca and Padua, being offered honorary PhD degrees by their universities in honor of Galileo, being the 350th year since his death.

The second of these memorable trips was to Padua in December 1992. The University of Padua was celebrating the inaugural lecture in 1592 of Galileo Galilei. It was at Padua that Galileo went on to use a telescope of his own design to study the sky, and discovered phases of Venus whose details decisively contradicted the geocentric theories of Aristotle and Ptolemy. With his telescope, Galileo found mountains on the Moon and measured their height, and observed thousands of stars that no one before him had ever seen. His discovery of the moons of Jupiter came later, after he moved to Florence.

My talk at the scientific and historical meeting on December 6 was about Galileo's later work on mechanics, work done when he was under house arrest at Arcetri. I emphasized Galileo's role in introducing time into mechanics, which had previously dealt mostly with statics and with the shapes of trajectories.

In 1993, I was interviewed on the Charlie Rose show. I much enjoyed it and appreciated him. This interview, among other interviews of me, I find appears on YouTube.

In 1994, we would find ourselves once more in Copenhagen, in December, revisiting the Bohr Institute and staying at our old haunt, the Hotel d'Angleterre in Copenhagen. The occasion was to honor the memory of Gunnar Källén, and I was invited to give the talk.

In March 1994, we were off to New York for a conference run by *Scientific American*. On that occasion, I recall a most memorable lunch with my agent, Mort Janklow, at the Chantilly. An intriguing and knowledgeable companion.

One day, I was at the blackboard describing the effective field theory of low-energy pions and nucleons. I described the rule that I had derived in 1979, that at low energy the dominant interactions that need to be taken into account are those in which the number of rates of change plus half the number of nucleon fields equals two with any number of pion fields, as required by the symmetries of the theory. For instance, there is an interaction with no nucleon fields and two rates of change of pion fields, which produces the scattering of pions by pions, and an interaction with two nucleon fields and one rate of change of a pion field, which produces the scattering of pions by nucleons. Suddenly at the blackboard, I realized that there was another kind of interaction that would satisfy this rule, an interaction with four nucleon fields and no rates of change or pion fields. It had taken me a decade to realize that half of four is two. This would produce reactions in which two colliding nucleons are destroyed at the point of collision and then created again. The exciting thing was that this was just the sort of point nucleon–nucleon interaction that nuclear theorists had known for years is needed, along with the exchange of pions between nucleons, to account for the known properties of nuclear forces. I saw the possibility of a systematic theory of nuclear forces at low energy, based on the chiral symmetry that governs pions and nucleons and the general ideas of effective field theories. I wrote a few papers along these lines in the next few years, but the real work of carrying these ideas forward has been done by other theorists, including Carlos Ordoñez and Bira van Kolck, who had been graduate students at Austin.

The 1990s and the century wound down with some developments worth noting briefly. In January of that year, Louise and I were off to the World Economic Forum at Davos, where we were each to serve as Forum Fellows. We were each assigned to chair a table discussion or two. A curious feature of the Davos meeting is that there is a tense atmosphere. I do not want to identify anti-Semitism. But there was a disinclination that was understood by others before we suspected it. The Israeli ambassador would come to one of my table talks at Davos, as if to support me; and it turned out to be helpful that he had. We had a good "power couples" table, where we had the company of dear friends Abe and Toni Chayes. But it was a table of Jewish power couples. Other tables were judenrein. In the late afternoons, we escaped to a delightful old cafe.

Through the early 1990s at the University of Texas at Austin (UT), I taught a course on the quantum theory of fields. As had happened earlier with my course on general relativity, over time my lecture notes became fuller, words began to appear among the equations, and the notes turned into the first draft of a book. I planned a two-volume treatise, *The Quantum Theory of Fields*. The first volume, *Foundations*, would explain the fundamentals of the subject. This would be done, not as usual, by applying widely used quantum mechanical assumptions to existing field theories, like Maxwell's theory of electrodynamics, but rather by deriving quantum field theories as a natural consequence of symmetry principles, including the spacetime symmetry underlying special relativity, and the general principles of quantum mechanics. The second volume, *Modern Applications*, would describe our modern theory of elementary particles, the Standard Model, as an application of quantum field theory. Much of this would involve material I had worked on myself, especially including the breakdown of symmetries, but some would involve arcana called anomalies and instantons, about which I was just recently learning.

In 1995 and 1996, I published two volumes on *The Quantum Theory of Fields*. These had good reviews. At a JASON meeting, Murph Goldberger paid me the compliment of saying that he liked my book because it expressed "a point of view." He couldn't have understood me better.

With the urgings of my Texas colleague, Jacques Distler, I wrote a third volume, *Supersymmetry*, published in 2000. None of the squarks, gluinos, and so on that supersymmetry puts into families with known particles of different spin had been discovered (and still have not been), but the theory is mathematically beautiful and offered at least a chance of explaining the huge disparity between the scale of masses of known particles and the vastly larger masses like the Planck mass that we suspect are relevant to a fundamental theory underlying the Standard Model. I wrote this third volume to ready myself and my readers for the happy day when future experiments uncover definite signs of supersymmetry.

I signed up for publication with Cambridge University Press. They are a classic publisher of physics treatises; they offered me a good cash advance; and most important my editor, Rufus Neal, was someone with whom I could work comfortably.

At this time, I began publishing popular science in *The New York Review of Books (NYRB)*. Robert Silvers, the founding editor, had a tolerance for me even when I may have been a bit too abstruse, and I remain grateful to him for the hospitality of the *NYRB*. Freeman Dyson wrote far more elegantly than me for that journal. I much appreciated Dyson's critique published there of Great Britain's Bomber Command in World War II.

Sometime that summer break, we were off to Barcelona, where Quim and Montse Gomis hosted and shepherded us. How hospitable the Gomises are, and how beautiful the city is, particularly with its new marine landscape.

On this trip, I gave a talk in Milan as well. The great Cathedral of Milan in its majestic square was a thrilling sight, with birds wheeling round its domes and pinnacles. The Mayor, Giuliano Pisapia, with his suave aristocratic aura, hosted us graciously.

In the mid-1990s, the Hobby–Eberly Telescope was launched here in Texas, at the McDonald Observatory. I was invited to deliver the dedicatory address on the occasion of its opening. Our good friends Jane and D. J. Sibley invited us to spend that night at the

Castle, their home on their nearby Glass Mountain ranch. After dinner, we sat outside and gazed up at the stars. I had never seen such stars. When I exclaimed at the dazzling display, D. J. complained that a Mexican factory was ruining their night sky. I wished I had seen it before the alleged ruining, but as it was, it inspired awe.

The Hobby–Eberly is run by UT, but is backed as well by other American universities. This amazing instrument is one of the world's largest optical telescopes, clearly a prototype of the James Webb, the next-generation space telescope being built as I write this. The main mirror of the Hobby–Eberly is made up of ninety-one hexagonal pieces fitted together, and, sitting at a 55-degree angle, it can be rotated around its base, and cover almost the entire sky from horizon to horizon. Three spectrographs work with the telescope to analyze the light from the phenomena under observation.

The Hobby–Eberly Telescope has discovered planets circling other stars. It has identified supernovas and has measured the rotation

FIGURE 17.1 The Hobby–Eberly Telescope

FIGURE 17.2 The author's home office in Austin

of particular galaxies. Further improvements have enabled it to join the so-called Dark Energy Experiment, and to explore black holes. It also is in a program to find other habitable planets, the implied dream of my anthropic argument. (I have argued that the universe must be understood as limited by the constraint that it must make possible a sun like ours and a planet like ours, since in fact ours does, and a planet that holds water, where a scientist can live and study it.) This anthropic argument is criticized by some physicists as insufficiently mathematical, but it provides a calculable measure of the dark energy. This is a calculation that is, in effect, Einstein's "cosmological constant," which he inserted for anthropic reasons. It is, after all, the force needed to prevent the gravitational collapse of the galaxies toward each other, and also prevent the uncontrolled flight of the galaxies apart from each other. Alas, the dark energy of space is in fact moving the galaxies apart from each other, slightly faster with every second of time.

In 1999, I was awarded the Lewis Thomas Prize for Writing about Science at Rockefeller University. We were dined in the President's glasshouse in the East River.

That December, we were off to Stockholm and the Grand Hotel for Nobel Week. We much enjoyed these nostalgic return trips. On this occasion, Gerhard 't Hooft and his advisor Martinus Veltman were the medalists, having shown that my electroweak theory was renormalizable, and I had recommended them for the prize, thus earning the return invitation.

Louise will forever twit me about all that. When we are having some little disagreement, she will say, "You couldn't even renormalize your own theory." I broke the thread of my summary talk to tell this story at a convocation at Case Western Reserve University in Ohio, celebrating "Fifty Years of the Standard Model." The audience of friends and colleagues got a belly laugh out of it.

Image Credits

Figure 2.1   Wikipedia image
Figure 3.1   Wikipedia image (Jerrye & Roy Klotz) CC BY-SA 4.0
Figure 3.2   Portrait by and image courtesy of Louise Weinberg
Figure 4.1   AIP Emilio Segrè Visual Archives [Weinberg A4]
Figure 5.1   AIP Emilio Segrè Visual Archives, Physics Today Collection [Wigner Eugene A3]
Figure 5.2   AIP Emilio Segrè Visual Archives, Segrè Collection [Treiman Sam B3]
Figure 6.1   H.-Y. Chin via National Archives and Records Administration
Figure 6.2   AIP Emilio Segrè Visual Archives, Wheeler Collection [Wheeler John Archibald C23]
Figure 6.3   AIP Emilio Segrè Visual Archives, Physics Today Collection [Feinberg Gerald A2]
Figure 7.1   Robert M. Couto, courtesy of Lawrence Berkeley National Laboratory. © The Regents of the University of California, Lawrence Berkeley National Laboratory
Figure 7.2   Courtesy of Louise Weinberg
Figure 7.3   AIP Emilio Segrè Visual Archives, Marshak Collection [Gell-Mann Murray C1]
Figure 8.1   AIP Emilio Segrè Visual Archives, Marshak Collection [Salam Abdus B1]
Figure 9.1   AIP Emilio Segrè Visual Archives, Physics Today Collection [Coleman Sidney B1]
Figure 9.2   Harvard University, courtesy of AIP Emilio Segrè Visual Archives, Weber Collection, Physics Today Collection [Schwinger Julian A3]
Figure 10.1 AIP Emilio Segrè Visual Archives, Physics Today Collection [Weinberg Steven B3]
Figure 10.2 Keystone-France/Gamma-Keystone via Getty Images [Image # 107708839]

Figure 10.3   Courtesy of International Solvay Institutes, Brussels
Figure 11.1   Jochmann Disco, Utrecht, courtesy of AIP Emilio Segrè Visual Archives, Physics Today Collection [Hooft Gerardus 't A4]
Figure 12.1   Gerald Present, Fermi National Accelerator Laboratory, courtesy AIP Emilio Segrè Visual Archives, Physics Today Collection [Lee Benjamin B1]
Figure 12.2   AIP Emilio Segrè Visual Archives [Weinberg Steven C4]
Figure 12.3   Associated Press/Alamy Stock Photo [Image # 2NCPDP9]
Figure 12.4   Associated Press/Alamy Stock Photo [Image # 2NFR1H9]
Figure 12.5   Bettmann via Getty Images [Image # 515123744]
Figure 13.1   AIP Emilio Segrè Visual Archives, Segrè Collection [Chew Geoffrey C3]
Figure 13.2   Sam Treiman, courtesy of AIP Emilio Segrè Visual Archives, Physics Today Collection [Weinberg Steven D7]
Figure 14.1   Mary Anne Fackelman-Miner, The White House, courtesy of AIP Emilio Segrè Visual Archives [Weinberg Steven D3]
Figure 14.2   AIP Emilio Segrè Visual Archives, Physics Today Collection [Weinberg Steven B14]
Figure 17.1   Marty Harris, McDonald Observatory, courtesy of the University of Texas
Figure 17.2   Courtesy of Louise Weinberg

Bibliography

Bernstein, J. *Oppenheimer: Portrait of an Enigma*, University of Chicago Press (2004)

Chandrasekhar, S. *An Introduction to the Study of Stellar Structure*, University of Chicago Press (1939)

Chandrasekhar, S. *Radiative Transfer*, Oxford University Press (1950)

Chayes, A. and Wiesner, J. B. (eds.) "A. B. M.: An Evaluation of the Decision to Deploy an Antiballistic Missile System," Harper & Row (1970)

Fadiman, C. (ed.) *Living Philosophies*, Doubleday (1990)

Feynman, R. P. and Weinberg, S. *Elementary Particles and the Laws of Physics*, Cambridge University Press (1988)

Freeman, D. S. *Lee's Lieutenants*, Charles Scribner's Sons (1942)

Galbraith, J. K. *The Affluent Society*, Houghton Mifflin (1958)

Gamow, G. "Mr Tompkins in Wonderland," *Discovery Magazine* (1940)

Gamow, G. *The Birth and Death of the Sun*, The Viking Press (1940)

Gamow, G. *Mr Tompkins Explores the Atom*, Cambridge University Press (1945)

Gamow, G. *One, Two, Three, ... Infinity*, The Viking Press (1947)

Goudsmit, S. *Alsos*, Henry Schuman (1947)

Hawking, S. W. *A Brief History of Time*, Bantam Books (1988)

Hawking, S. W. and Israel, W. *General Relativity: An Einstein Centenary Survey*, Cambridge University Press (1979)

Heitler, W. *Quantum Theory of Radiation*, Clarendon Press (3rd ed., 1954)

Hitchcock, A. (dir.) *The Lady Vanishes*, (movie) Gainsborough Pictures (1938)

Jeans, J. *The Mysterious Universe*, Cambridge University Press (1931)

Keary, A. and Keary, E. *The Heroes of Asgard*, The MacMillan (1893)

Kurti, N and Kurti, G. (eds.) *But the Crackling is Superb*, Adam Hilger (1988)

Misner, C., Thorne, K. and Wheeler, J. "Gravitation," W. H. Freeman & Co. (1973)

Newton, I. *Philosophiæ Naturalis Principia Mathematica*, (1687)

Peebles, P. J. E. "Large-Scale Structure of the Universe," Princeton University Press (1980)

Schiff, L. *Quantum Mechanics*, McGraw-Hill Book (1949)

Sommerfeld, A. *Electrodynamics*, Academic Press (1952)

Weinberg, S. *Gravitation and Cosmology*, John Wiley (1972)

Weinberg, S. *The First Three Minutes*, Basic Books (1977)
Weinberg, S. *The Discovery of Subatomic Particles: Revised Edition*, Scientific American Library (1983), Cambridge University Press (2003)
Weinberg, S. *Dreams of a Final Theory*, Hutchinson Radius/Vintage (1993)
Weinberg, S. *The Quantum Theory of Fields: Vol. 1. Foundations*, Cambridge University Press (1995)
Weinberg, S. *The Quantum Theory of Fields: Vol. 2. Modern Applications*, Cambridge University Press (1996)
Weinberg, S. *The Quantum Theory of Fields: Vol. 3. Supersymmetry*, Cambridge University Press (2000)
Weinberg, S. *To Explain the World*, Penguin (2016)
White, T. H. *The Once and Future King*, Collins (1958)
Whittaker, E. T. and Watson, G. N. *A Course of Modern Analysis*, Cambridge University Press (1902)
Yeats-Brown, F. *The Lives of a Bengal Lancer*, The Viking Press (1930)

Index

accelerators, case for, 210
All Souls College (Oxford), 193
alpha decay, 19
American Academy of Arts and Sciences, election to, 107
Anderson, Philip, 14, 220, 228
anthropic principle, 227, 228
Anti-Ballistic Missile (ABM) program, 109, 113
antimatter, 79, 81
Artin, Emil, 28
asymptotic freedom, 130, 134, 140
asymptotically safe theories, 157
   gravitation, 158
Austin, Texas, 159, 176
   ice-skating, 179
axion, 150

Bardeen, John, 14
Bargmann, Valentine, 26
baryon number, 165
Baryshnikov, Mikhail, 159
beauty in physics theories, 207, 228
Beijing, 155
Bell Laboratories, 13, 15
Bernstein, Jeremy, 51
beta decay, 31, 32, 36, 42
Bethe, Hans, 15
Bjorken scaling, 130
Bobbitt, Philip, 160
Bohr, Niels, 18
   and Margarethe, 23
   and superstition, 23
book reviews, 123
   writing, 89
   Wigner, 113
book writing
   Dreams of a Final Theory, 225
   The Quantum Theory of Fields, 25, 241
   Gravitation and Cosmology, 75, 114, 232

The Discovery of Subatomic Particles, 67, 178
The First Three Minutes, 135, 145
Brézin, Édouard, 156
broken symmetries, 57
Bronx High School of Science, 5
Brueckner, Keith, 49

Cabibbo, Nicola, 68
   Cabibbo angle, 69
Caianiello, Eduardo, 68
Cal Tech
   apprehension of, 52, 87
Canary Islands visit, 203
Carter, Jimmy (US President), 152
Cavendish Laboratory, Cambridge, 67
CERN
   LHC, *see* Large Hadron Collider
   Proton Synchrotron, 69
   visit to, 69
Chandrasekhar, Subrahmanyan, 45
Chang, Lay Nam, 106
Chayes, Abram, 109
Chew, Geoffrey, 43, 48, 181
Clare College, Cambridge, 89
Coleman, Sidney, 85
   on Princeton, 31
Collège de France, 116
Columbia University, 35
   Chinese lunches, 38, 42
   leaving, 44
committee work, 111, 149, 152, 193
Copenhagen (play), 18
Coral Gables meeting, 125
Cornell University, 7, 8
cosmic microwave background, 114, 232
cosmological constant problem, 227, 234
Coyne, George, 179
critical phenomena, 142
CTP theorem, 70
current algebra, 90, 94, 111, 141

250 INDEX

Dalitz, Richard, 16
Dannie Heineman Prize, 151
dark energy, 230, 235
    supernovae observations, 230
dark matter, 148
    axions, 151
    WIMPs, 148
Dirac lectures, 206
Dirac theory (quantum mechanics), 77
Dirac, Paul, 77, 80
Dyson, Freeman, 28, 29

Eddington, Arthur, 147
effective field theory, 33, 93, 164, 206
Einstein, Albert, 12
    cosmological constant, 113
electroweak theory, 100, 101, 116, 119, 122, 126
    coupling constants, 138
    discovery of W and Z particles, 200
    experimental verification, 129, 144, 152
    renormalization of, 104
Erice summer school, 143
Escuela Normale Superiore, Pisa, 69

Faddeev, Ludwig, 122
Fadiman, Clifton, 237
Feinberg, Gerald (Gary), 6, 7, 39, 43, 51, 216
Feld, Bernie, 112
Feynman, Richard, 9, 87, 197
    and electroweak theory, 129
    path integral formalism, 27
Fischler, Willy, 198
Foley, Henry, 44
Frisch, David, 18
Fubini, Sergio, 56
functional analysis, 75
    work of Faddeev, 76
fundamental theories, 83

Galileo Galilei, 238
Gamow, George, 3
gauge invariance, 82
gauge transformation, 82
Gell-Mann, Murray, 40, 41, 106
    and astrophysics, 115
    Gatlinburg Conference, 40
    quark theory, 130
Georgi, Howard, 138
Glaser, Don, 47

Key West incident, 53
Glashow, Sheldon (Shelly), 6, 7, 10, 85
Goldberger, Marvin (Murph), 30, 43, 126
Goldberger–Treiman relation, 58
Goldstone bosons, 58, 150
Goldstone, Jeffrey, 56
Gore, Al, 223
Griffy, Tom, 183

Harvard University
    Higgins Professorship, 124
    Loeb Lectureship, 88
    resignation from, 185
Harvard-Smithsonian Center for Astrophysics, 124, 135
Headliners Club, Austin, 185
Heisenberg, Werner, 102
Herglotz theorem, 52
hierarchy problem, 139
Higgs boson, 100, 117, 210
    discovery, 225
Higgs, Peter, 67
Hobby, Bill, 184
Hobby–Eberly Telescope, 184, 241
Holton, Gerald, 167
Hong-Mo, Chan, 61
honorary degrees, 154, 159, 175, 202, 237, 238
Hoyle, Fred, 115
Huang, Kerson, 115, 235
hyperphoton, 86

interferometry, 72
International Centre for Theoretical Physics (Trieste), 83

J particle, 139
Jackiw, Roman, 121
JASON, 49
    end of involvement, 123
    HBT on a Wake, 72
    joining, 49
    Key West exercise, 53
    magnetohydrodynamics, 58
    tactical nuclear weapon study, 95
    US intelligence, 84
Jerusalem, 188
Jerusalem Winter School in Theoretical Physics, 64, 188, 204

K meson, 47, 139

INDEX   251

Kabir, Pasha, 38
Kac, Mark, 13, 17
Källén, Gunnar, 19, 21, 22
Kaluza–Klein theory, 192
Khuri, Nicola (Nick), 33, 195
Kibble, Tom, 64, 99
Kleitman, Danny, 9
Kroll, Norman, 42
Krumhansl, James, 220
Kuhn, Thomas, 8

Lamb, Willis, 194
Large Hadron Collider (LHC), 211, 222
Lee model, 21, 23, 25, 31, 35
Lee, Ben, 119, 147
    premature death, 149
Lee, Tsung-Dao (T. D.), 21, 34, 35, 73, 101, 155
lepton number, 165
leptons, model of, 116
    article, 100
Lewis Thomas Prize, 244
Linde, Andrei, 231
Los Alamos
    misdirection, 42
    visit to, 59
loss of notes incident, Singapore, 62
Low, Francis, 68
    Gell-Mann–Low paper, 68

MIT
    arrival, 97
    departure from, 124
Mach, Ernst, 30
Mandelstam, Stanley, 42, 43
Marshak, Robert, 41, 50
Martians, the, 48
Martin, Paul, 124
massless particles, 81, 83, 86
McIntosh, Harold, 13
Minkowski, Hermann, 12
mulecule, 47
multiverse, 228, 230
muon, 47

Nambu, Yochiro, 57
National Academy of Sciences
    election to, 123
National Medal of Science, 237
Ne'eman, Yuval, 182, 187
neutrino problem, 55

neutrinos, 135
    in cosmology, 148
    mass of, 165
Niels Bohr Institute, 16, 18
Nobel Prize
    announcement, 166
    ceremony, 169
    ditty, 81
    prediction, 142

Olbers' paradox, 53
Oppenheimer Prize, 125
Oppenheimer, J. Robert, 4, 15, 30, 42, 70,
    122, 194
    and Kitty, 122

Pal, Yash, 62
Parker, Eugene, 49
Pauli, Wolfgang, 21, 22, 25
Peccei, Roberto, 149
Peccei–Quinn symmetry, 150
Penzias, Arno, 15, 114
Perrin, Jean and Francis, 102
personal computers, 208
perturbation theory, 65
Philosophical Society of Texas, 198
    presidency, 232
pion scattering, 91, 94
Piran, Tsvi, 187
Polchinski, Joe, 198
Pomeranchuk, I. Ya, 52
positron, 80
Press, Frank, 152
Princeton University
    Atomic Beam Group, 24
    research ethos, 25

quantum chromodynamics, 131
quantum electrodynamics, 28
    Feynman, Schwinger, and Tomonaga, 20
    Källén, 19
quantum theory of fields, 76
    canonical formalism, 76
    particle formalism, 79
quarks, 130
    color trapping, 132
    masses, 146
Quinn, Helen, 138, 149

Rabi, I. I., 37

252   INDEX

Rathjens, George, 109
Reagan, Ronald (US President), 212
reductionism, 207
renormalization, 20, 38, 101, 157
Rochester Conference
   Berkeley (1966), 92
   Berkeley (1986), 205
   Dallas (1992), 238
   Dubna (1964), 84
   Geneva (1962), 69
   Vienna (1968), 106
round-the-world trip, 56
Royal Society of London, 209
   cookbook, 209
Rubbia, Carlo, 37, 127, 200
Rutherford, Ernest, 19
   Cavendish crocodile, 67

Safeguard. *See* Anti-Ballistic Missile (ABM) program
Salam, Abdus, 29, 48, 64, 169
Saturday Club, Boston, 174
Schmidt, Maarten, 136
Schwartz, Melvin (Mel), 36
Schwinger, Julian, 37, 88
   and Clarisse, 89
   festschrift, 162
   shoes, 126
Schwitters, Roy, 216
science fiction, 4
Segrè, Gino, 11
Sentinel. *See* Anti-Ballistic Missile (ABM) program
Serber, Robert (Bob)
   and Charlotte, 42
Shannon, Claude, 14
Shapiro, Paul, 229
Shelter Island Conference
   1947, 374
   1983, 375
Sirlin, Alberto, 42
S-matrix theory, 48, 71
Solvay Conference
   1911, 349
   1967, 348
   photographs, 102, 103, 105
spectral function sum rules, 94
spin, 173
   in quantum mechanics, 77
Standard Model (of particle physics), 116, 168

   birth of, 122
Stanford University, 145
steady-state universe, 115
Steinberger, Jack, 37
Stern, Isaac, 202
Stockholm
   1979, 367
   1983, 369
   1988, 370
   1991, 371
   1999, 368
string theory, 203
Sudarshan, George, 41
Superconducting Super Collider (SSC), 117, 211
   cost, 221
   demise of, 224
   politics, 218
   site selection, 214, 215
supersymmetry, 190, 210
Susskind, Leonard (Lenny), 154
symmetry breaking, 67, 73, 100

teaching
   general relativity, 74, 75, 113
   quantum theory of fields, 240
   subatomic particles, 178
   vector algebra, 50
technicolor coupling, 154
't Hooft, Gerard, 116, 149, 172
Telegdi, Valentine, 47
Telluride House (fraternity), 9, 13
Tomonaga, Sin-Itiro, 60
Town and Gown Club, Austin, 200
Treiman, Sam, 30, 72
   and Joanna, 31, 73
   as thesis advisor, 33
Trinity College, Cambridge, 230
Tuesday Club, Austin, 201

UC Berkeley
   assistant professorship, 49
   associate professorship, 71
   Bevatron, 46, 81
   full professorship, 88
   Radiation Laboratory, 44, 45
   resignation from, 112
   slipped disc, 45
Umezawa, Hiroomi, 60
Union of Concerned Scientists, 110
University of Texas at Austin

moving to, 184
Theory Group, 198

Vatican, the, 179
Veltman, Martinus, 172
Vienna, 107

W particle, 39, 80
Watson, Kenneth, 59
weak interactions, 31, 38, 51
    at Rochester, 41
Weinberg, Louise, 16
    application to law school, 88
    née Goldwasser, 9
    professorship, 198
    University of Texas School of Law, 158, 160
Wheeler, John Archibald, 27, 59
Wiener, Norbert, 68
Wiesner, Jerome, 109
Wightman, Arthur, 27
Wigner, Eugene, 25, 113
Wilczek, Frank, 150, 185
Wilets, Larry, 63
Wilson, Robert, 15, 114
Witten, Edward, 173
World Economic Forum, Davos, 240
Wu, Chien-Shung, 36

Yang–Mills theories, 130

Z particle, 100, 120
Zichichi, Antonino, 143
Zumino, Bruno, 70
    and Mary Gaillard, 172
    and Shirley, 70, 107
Zweig, George, 130

Def of vectors / rotatn ref frame
Vectors - 51 $\vec{r'}$ — 96 vectrs in R⁴

Symmetry 56

Books - p.67 = "Discov / Subatomic Particles"
p.75 = "Gravitatn + cosmology"